INDUSTRIAL SAFETY AND
RISK MANAGEMENT

INDUSTRIAL SAFETY AND RISK MANAGEMENT

INDUSTRIAL SAFETY AND LOSS
MANAGEMENT PROGRAM
FACULTY OF ENGINEERING
UNIVERSITY OF ALBERTA

Laird Wilson, P.Eng.

CO-AUTHOR
Doug McCutcheon, P.Eng.

CONTRIBUTORS
Marilyn Buchanan
Davis Bourque
Dieter Brunsch
Daneve McAffer
Gerry Phillips, P.Eng.
Don Ritz
Ron Rosser, P.Eng.

The University of Alberta Press

Published by

The University of Alberta Press
Ring House 2
Edmonton, Alberta, Canada T6G 2E1

Copyright © Industrial Safety and Loss Management Program, Faculty of Engineering, University of Alberta 2003

ISBN 0-88864-394-2

**National Library of Canada
Cataloguing in Publication Data**

Wilson, Laird, 1929–
  Industrial safety and risk management / Laird Wilson ; co-author, Doug McCutcheon ; contributors, Marilyn Buchanan ... [et al.].

  "Faculty of Engineering, University of Alberta, Industrial Safety and Loss Management Program".
  Includes bibliographical references.
  ISBN 0-88864-394-2

  1. Industrial safety. 2. Risk management.
I. McCutcheon, Doug. II. Buchanan, Marilyn.
III. University of Alberta. Faculty of Engineering.
Industrial Safety & Loss Management Program.
IV. Title.
HD7261.W55 2003   363.11   C2003-911179-2

First edition, first printing 2003

Printed and bound in Canada by Houghton Boston Printers and Lithographers Ltd., Saskatoon, Saskatchewan.

The University of Alberta Press is committed to protecting our natural environment. As part of our efforts, this book is printed on stock produced by New Leaf Paper: it contains 100% post-consumer recycled fibres and is acid- and chlorine-free.

The University of Alberta Press gratefully acknowledges the financial support of the Government of Canada through the Book Publishing Industry Development Program (BPDIP) and from the Alberta Foundation for the Arts for our publishing activities.

THE CANADA COUNCIL | LE CONSEIL DES ARTS
FOR THE ARTS | DU CANADA
SINCE 1957 | DEPUIS 1957

CONTENTS

Figures and Tables   viii

Foreword by JAMES E. CARTER   ix

Preface   xi

Acknowledgements   xiii

Definitions and Terminology   xv

INTRODUCTION   1

1   Industrial Health, Safety and Risk Management— The Integrated Approach   3

■ *Introduction* ■ *Historical Background* ■ *The Integrated Approach and Benefits—PEAP* ■ *Company Policy Statement and Policies* ■ *Value Systems and Cost/Benefit* ■ *Summary*

REGULATIONS AND PROFESSIONAL RESPONSIBILITIES   9

2   Government Regulations   *DIETER BRUNSCH and DAVIS BOURQUE*   11

■ *Federal and Provincial Legislation* ■ *Other Countries Systems and Legislation* ■ *Alberta Legislation*

3   Due Diligence   *DAVIS BOURQUE*   15

■ *Definition of Due Diligence* ■ *Summary*

4   Professional Organizations and Professional Responsibilities   21

■ *Professional Organizations* ■ *Benefits to Professionals, Industry and the Public* ■ *Code of Ethics* ■ *Responsibility Under Law* ■ *Reduction of Losses and Human Suffering* ■ *Benefits to the Professional and the Company*

INDUSTRIAL SAFETY AND RISK MANAGEMENT PROGRAMS   27

5   Industrial Health, Safety and Risk Management Programs   29

■ *Introduction* ■ *A Typical Industrial Health, Safety and Risk Management Program* ■ *Successful Program Implementation* ■ *Cost Effectiveness*

6   Safety and Risk Management for Project Managers   39

■ *Introduction* ■ *The Integrated Approach to Risk and Loss Reduction* ■ *The Importance of Safety and Risk Management for Project Managers* ■ *The Role of Project Managers* ■ *Project Managers' Areas of Responsibility* ■ *Steps to Achieve Excellence by Company X* ■ *Summary*

7   Risk Assessment, Analysis and Management   *GERRY PHILLIPS*   43

■ *The Risk Management Process* ■ *Hazard Identification and Risk Assessment Techniques* ■ *Summary of Methodologies*

8   Causation Model and the Importance of Systematic Incident Investigation   59

■ *Immediate Causes* ■ *Basic Causes* ■ *Lack of Control (Stewardship)* ■ *Importance of Systematic Incident Investigation* ■ *Fragility of Evidence* ■ *Interviewing Techniques and Helpful Hints* ■ *Root Cause Analysis* ■ *Summary*

9   Human Factors   *DON RITZ*   67

■ *Introduction* ■ *What Are Human Factors?* ■ *Case Study—Application of Human Factors to a Project* ■ *Case Study—Information Display Improvements in a Control Centre* ■ *Case Study—Workplace Ergonomics Program in Operations* ■ *Case Study—Human Factors Application by a Plant Maintenance Group* ■ *Human Factors Standards* ■ *Case Study—Noise Standards*

CASE STUDIES   83

10   Causes of Industrial Disasters and Lessons Learned   85

■ *The Flixborough Disaster* ■ *Piper Alpha* ■ *Challenger Space Shuttle Disaster* ■ *Hyatt Regency, Kansas City* ■ *The Titanic*

11   Case Studies of Canadian Industrial Incidences   93

■ *Introduction* ■ *Lodgepole* ■ *Syncrude Coker 8-2 Fire* ■ *References*

SPECIFIC TOPICS   107

12   Contractors   109

■ *Introduction* ■ *Historical Background* ■ *Creation of an Effective Relationship Between Contractors and Companies* ■ *Ongoing Evaluation of the Contractor* ■ *APPENDIX 12–1: Construction Owners Association of Alberta*

13   Small Company Performance and Program   117

■ *Introduction* ■ *Strategies for Small Companies* ■ *Sociological Study on Small Businesses* ■ *Ideas for Improvement* ■ *Occupational Health and Safety Manual for Small Businesses*

14 Risk Communication and Public Emergency Response — 123

■ *Risk Communication and Community Awareness* ■ *Emergency Response*

15 Safety and Risk Management for Young Workers
with MARILYN BUCHANAN — 127

■ *Introduction* ■ *Causes* ■ *Training Young Workers to Avoid
At-Risk Behaviour*

16 Provincial Regulatory Agencies and
Workers' Compensation Boards *DIETER BRUNSCH* — 131

■ *Provincial Regulatory Agencies* ■ *Workers' Compensation Boards*
■ *WCB-Alberta Vision* ■ *Summary* ■ *APPENDIX 16–1: Canadian Federal Acts Around
Safety and Health* ■ *APPENDIX 16–2: OH&S Legislation in Canada—Basic
Responsibilities* ■ *APPENDIX 16–3: Understanding Workplace Health and Safety*

17 Key Benefits of Computer Systems
and Communications *RON ROSSER* — 145

■ *Process Controls* ■ *Management of Software and Hardware*
■ *Statistical Tools and Administrative Systems for Managing
Results* ■ *Typical Administrative Systems*

BUILDING A PROGRAM — 149

18 Desired Safety and Risk Management Results
Through Team Empowerment *DANEVE McAFFER* — 151

■ *High Performing Teams* ■ *Changing Attitudes* ■ *Organizational
Effectiveness* ■ *Enhancing Team Relationships* ■ *Conclusion*

19 Implementing an Industrial Safety
and Risk Management Program — 159

■ *Developing a Training Program* ■ *The Benefits of a First-class Program*
■ *Importance of Industrial Risk Management to Companies and Their
Employees* ■ *Importance of Senior Management Involvement* ■ *Conclusion*

Afterword by NORM MCINTYRE — 163
References — 171

Figures and Tables

Figure 1–1: The Definition of Industrial Safety and Risk Management.   6

Figure 1–2: An Example of a Policy Statement.   7

Figure 4–1: The Safety Cycle.   24

Figure 4–2: Effective Application for First-class Performance.   25

Figure 5–1: Risk Management Flow Chart.   31

Figure 5–2: Management System for Implementation of a
Safety and Risk Management Program.   36

Figure 5–3: First-class Company Results.   38

Figure 7–1: The Risk Management Process.   44

Figure 7–2: Types of Risk Assessment/Risk Analysis Methods vs. Time Invested.   48

Figure 7–3: Simplified Logic Tree Analysis for an Explosion.   49

Figure 7–4: Fault Tree Diagram with AND and OR Logic Gates.   56

Figure 8–1: Incident Causation Model.   63

Figure 8–2: Incident Investigation Flow Path.   64

Figure 8–3: Preservation of Evidence.   65

Figure 9–1: Human Factors Model.   71

Figure 10–1: Layout of Reactors and Temporary By-pass at Flixborough.   86

Table 7–1:   Semi-quantitative Risk Assessment.   51

Table 7–2:   "Generic" Risk Criteria.   52

Table 7–3:   FMECA Worksheet.   54

Table 7–4:   The Guideword Approach.   55

Table 7–5:   Comparison of Analysis Methods.   57

Table 8–1:   Examples of Immediate Causes.   60

Table 8–2:   Injury Causation Table with Examples and Comments.   61

Table 8–3:   Examples of Basic Causes.   62

Table 9–1:   The Human Factors Model.   70

Table 9–2:   Noise Levels Associated with Various Locations.   76

Table 9–3:   Suggested Maximum Noise Levels within Various Locations.   77

Table 9–4:   Alberta Occupational Exposure Limits for Noise.   78

Table 9–5:   Alberta Occupational Exposure Limits (Impulse Noise).   79

Table 9–6:   Checklist for Noise.   80

Table 12–1: Example of a Policy Statement of Company A.   111

Foreword

James E. Carter
President and Chief Operating Officer
Syncrude Canada Ltd.

Laird Wilson and I both signed on for our adventures at Syncrude at about the same time. The sun was setting on the 1970s, rising on this very young oil sands outfit and rising also on the next phase of each of our careers in the oil sands industry. By playing a part in what Peter C. Newman would later call a *Canadian Success Story*, we realized the dream of many earlier generations of would-be oil sands entrepreneurs. They were the pioneers who had recognized the potential for wealth and opportunity in this vast resource, but who had not been able to discover the technical or commercial wherewithal to make it viable.

But ours was the generation when it came together. On the foundation of earlier efforts, we thought we could unlock the potential and achieve something truly great. And we have. Syncrude and the oil sands industry are today playing increasingly significant roles fueling the energy portfolio of an entire continent and, along the way, generating all kinds of economic and social opportunity for many thousands of Canadians.

But there is an equal, perhaps greater, measure of our success—indeed, of the success of any enterprise. It has to do with the good health, overall well-being, and prospects for those people directly or indirectly affected by what we do at Syncrude. Yes, we offer an *excellent* employment experience. But because it makes good business sense and reflects our commitment to doing the right thing, we also demand a *safe* employment experience. After all, what good is energy security and economic and social prosperity if we cannot share and enjoy the fruits of our labours in all aspects of our lives?

To me, the answer is self-evident. Although we likely cannot eliminate risk entirely, we can manage it effectively. And we have every reason to devote Herculean codes of conduct and guiding principles to our efforts. A human life—ours or someone else's—is priceless and we must do all we can to protect it. And not because we see a profit in doing so. But simply because it is the right thing to do. All businesses are, after all, made up of human beings and there is nothing wrong with factoring our humanity, a value-driven code of conduct, into the risk management proposition. As it turns out, though, there is a financial angle here as well. Effective management in this area saves the business money—insurance costs, for example—and results in a smoother, more reliably run operation.

Safety and risk management (or loss management as we tend to call it at Syncrude) are the tracks Laird took during his most influential years at Syncrude, and more lately in his academic career at the University of Alberta. It is also an issue that I have taken a keen and abiding interest in over my many years at Syncrude. To me, effective risk management is about doing all that is humanly possible to preserve the safety and well-being of yourself and those around you. It is about looking ahead, and for most readers of this book, that means learning all you can about safety and risk management in the relatively harmless—though not entirely risk free—environment of the classroom or company training situation.

For budding engineers, future business leaders, or anyone else coming in contact with any manner of industrial operation, *Industrial Safety and Risk Management* contains a wealth of valuable information touched by years of first-hand know-how and effective commitment to the preservation of life and limb in the workplace and beyond. It sets the stage for you in the years ahead to write your own success stories of prosperous careers and healthy and happy lives. As much as I encourage you to live and work safe in all your future endeavours, I commend this book to you with the confidence of my deep and abiding respect for Laird Wilson and co-author Doug McCutcheon.

Preface

The Industrial Safety and Loss Management Program (ISLMP) was designed at the request of Dr. Fred Otto, then Dean of Engineering at the University of Alberta in the late 1980s. Professor Laird Wilson, P.Eng., with over thirty years industrial experience, was appointed to the task of designing, starting up and continuously improving the overall program. It was Dr. Otto's insight into what the industry required of graduating engineers in the area of Safety and Loss/Risk Management that shaped this program, which commenced in 1988. The program at the University of Alberta was and still is very successful with the strong backing of Dean David Lynch. It is now directed and taught by Professor Doug McCutcheon, P.Eng.

Industrial Safety and Risk Management is based on Professor Wilson's experience and input from industry and government. Until now, no textbook was available that covers the complete range of the program. It is the intention of this textbook to provide information for students, teachers, industry personnel and appropriate government departments. It is based on the continuously improved ISLMP, plus additional material from friends of the program.

In this book we have covered the main theme of both ENGG 404 (Industrial Safety and Loss Management) and ENGG 406 (Industrial Safety and Risk Management) as given to fourth-year engineers and business students at the University of Alberta. There is certainly more material presented in the lectures so the student can have a wider perspective of the topic. The lectures for each course are printed and available from the

University of Alberta Bookstore for a nominal charge. These lecture notes provide further information for students and are updated each year. At present it is planned that this textbook will be updated every 3 years. Others interested in the lecture notes can contact the co-author at the University of Alberta at doug.mccutcheon@ualberta.ca.

Acknowledgements

The Industrial Advisory Committee provided valuable input to this book through the Industrial Safety and Loss/Risk Management program, Faculty of Engineering, University of Alberta. Members of the Advisory Committee are representatives of the following agencies, organizations and corporations:

A.P.E.G.G.A.
Alberta Environment
Alberta Federation of Labour
Alberta Human Resources and
    Employment
ATCO Electric
ATCO Gas
Celanese Canada Inc.
Chevron Canada Resources
C.N. Western Canada
C.P. Rail
Dow Chemical Canada Inc.
DuPont Canada Inc.
Imperial Oil Ltd.
NOVA Corporation of Alberta
PanCanadian Petroleum Ltd.

Petro-Canada Inc.
D.B. Robinson & Associates Ltd.
Shell Canada Ltd.
Suncor Inc. Oil Sands Group
Syncrude Canada Ltd.
Talisman Energy Inc.
Union Carbide Canada Inc.
University of Alberta
Weldwood of Canada Ltd.
Workers' Compensation Board, Alberta

More detailed input was provided to the appropriate chapters by the following:

Chapter 2	Dieter Brunsch, WCB, Davis Bourque, OH&S
Chapter 7	Gerry Phillips, NOVA Chemicals Ltd.
Chapter 9	Don Ritz, Syncrude Canada Ltd.
Chapter 15	Marilyn Buchanan, CRSP
Chapter 16	Dieter Brunsch, WCB

Chapter 17   Ron Rosser, Syncrude
            Canada Ltd.
Chapter 18   Daneve McAffer,
            Freeborn & Associates
            Consulting

The Syncrude Coker Fire case study in Chapter 11 is used with the permission of Syncrude. The Student Investigation Team who, under the guidance of Professor Wilson, prepared the original report included Scott Arndt, Allyson Belland, Ingrid Belle, Rod Godwaldt, Lance Hofer and Karl Norrena.

Appendix 12–1 Construction Owners Association of Alberta—COAA Safety Committee Mandate and Objectives For 2000–2001 is taken directly from their website and used with the permission of the Construction Owners Association of Alberta.

Chapter 15 by Marilyn Buchanan, CRSP, is based on research paper she completed as part of a Brain and Behaviour course at Red Deer College in social psychology.

Special thanks must be given to former Dean Fred Otto, and present Dean David Lynch, Faculty of Engineering, University of Alberta, for their strong backing and encouragement for this project.

Special thanks must also be given to Chanté van Rooyen, Administrative Assistant for the ISLM Program since 1993, for her patience, advice, hard work, dedication to the program and assistance in compiling the material for this publication.

Finally, we have received permission to reprint material from:

Alberta Energy and Utilities Board, Calgary, Alberta (Robert D. Heggie).

Canadian Centre for Occupational Health and Safety, Hamilton, Ontario (Roger Cockerline).

Canadian Standards Association, Toronto, Ontario (Lance Novak).

Coker Fire Incident, Syncrude.

Construction Owners Association of Alberta, Edmonton, Alberta (Brad Anderson).

Det Norske Veritas Inc., Loganville, Georgia (Bryan Robbins).

The Engineering Council, London, UK (Tony Miller, Senior Executive of Public Affairs).

Imperial Oil, Toronto, Ontario (Tony Pasteris).

Industrial Accident Prevention Association, Toronto, Ontario (Peter Nixon, Manager, Marketing and Communications).

Workplace Health and Safety and Employment Standards, Edmonton, Alberta (Wally Baer).

Full citations are included in the text.

Definitions and Terminology

As the area of Industrial Safety and Risk Management evolved, a different terminology was developed. In order to communicate within this topic, it is important to ensure that everybody uses the same terminology.

The list below is mainly derived from the Industrial Safety and Loss Management program at the University of Alberta, Faculty of Engineering, the Industrial Accident Prevention Association and the Canadian Standards Association. The first five definitions are highlighted since they are very important to understand in order to get the full benefit from this document. The remaining definitions are in alphabetical order.

INTEGRATED SAFETY AND RISK MANAGEMENT PROGRAM (ISRM)
A program designed to reduce the risk to people, environment, assets and production (PEAP) in an integrated manner in all industrial settings. Note this is what we mean when we use the term "safety and risk management program" in this document.

ACCEPTABLE RISK
A management tool used to set the criteria for what level of risk the company is willing to accept, i.e., risk matrix. It often involves public opinion and longer term issues. Regulations often define the basic level of acceptable risk and generally companies will set their standards to exceed the regulations in order to ensure they survive. Acceptable risk will be different for different companies; actually no two companies' policies are the same.

ACCIDENT
The connotative meaning of accident in the context of industrial safety and risk management refers to an undesired event that is believed to

be beyond the control of man. Since leading companies and loss prevention specialists believe all 'accidents' are preventable, it is argued that the term accident is not only wrong, but the use of the term accident is counterproductive to risk management initiates.

ACCIDENT RATIO
For every Lost Time Injury (LTI) there are 15 Medical Aid Injuries (MA), 150 First Aid Injuries (FA) and 600 Near Misses (NM). The major point is to work very hard at reducing the more numerous minor injuries and near misses. They provide a much larger basis for more effective control.

INCIDENT
An undesired event that does or could result in injury to people, damage to the environment or loss of assets and/or production. This includes both an actual loss or a near miss. An incident leading to a loss is most often the result of contact with a substance or a source of energy (mechanical, electrical, thermal, etc.) above **the threshold limit** of the body or structure involved or the environment.
Note: 'Incident' is a better term than 'accident', because accident has the implication of being caused by negative luck. The term 'incident' will be used in this text.

■ ■ ■ ■ ■

ASSESSMENT
A process that evaluates activities, facilities or systems against requirements or expectations.

BASIC CAUSES
The underlying or basic factors that allowed, and perhaps even invited, the immediate causes to develop. The reasons for the existence of immediate causes are substandard practices and conditions. They are more difficult to identify and are often not evident until after an incident has been thoroughly researched and investigated. These often describe deficiencies in the management system. Sometimes the term "root causes" is used.

COMPETENT PERSON
In the Occupational Health and Safety Act of Alberta, "competent" in relation to a worker means "adequately qualified," suitably trained and with sufficient experience, safety to perform work that is the subject matter of the relevant provision of this Regulation (the General Safety Regulation of the Act) without or with only a minimal degree of supervision.

CONTROLS
Procedures, processes or hardware (i.e., valves) that are put in place to prevent an incident from happening, to eliminate the probability where possible or to mitigate the incident to a safe level should it happen. Controls can include emergency plans and computer software.

CRITICAL FEW
A basic management principle that states that a small percentage of specific items account for the majority of all incidents and costs. (The 80/20 rule, as defined by Vilfredo Pareto (1843–1923).)

ELEMENTS
The elements of an Integrated Safety and Risk Management Program are:

> Management Leadership, Commitment and Accountability;
> Risk Assessment and Management;
> Design and Construction;
> Process and Facilities Information, Documentation;
> Personnel and Training;
> Operations and Maintenance;
> Management of Change;
> Contractors;
> Incident Investigation and Analysis;
> Community Awareness and Emergency Preparedness;
> Program for Continuous Improvement.

Each element is defined and has specific standards and objectives. They form the basis for designing, constructing and operating the company's facilities and for stewardship of performance.

FAULT TREE ANALYSIS
A systematic approach used in reliability analysis of complex systems in which the probabilities of failure of individual components and the resulting chains of cause-effect consequences are estimated. The method is frequently used to estimate the probability of a major accident.

FLASHPOINT
The lowest temperature at which vapours above a volatile combustible substance ignite in air when exposed to flame.

FREQUENCY RATE
Frequency rate = $\dfrac{(\text{number of injuries} \times 200{,}000 \text{ hours})}{(\text{total exposure hours})}$

Note: Total exposure hours = the number of persons x the number of hours worked. 200,000 hours represents the total approximate time that 100 persons would work in one year.

FIRST AID INJURY (F.A.)
Injury is attended to through the use of standard first aid treatments and there is no lost time from the job.

HAZARD
The potential of a machine, equipment, process, material or physical factor in the working environment to cause harm to people, environment, assets or production. For example, for a chemical it is the potential the substance has for causing adverse effects at various levels of exposure.

HAZARD IDENTIFICATION
The recognition that a hazard exists and the definition of its characteristics.

HOUSEKEEPING
A way of controlling hazards along the path between the source and the worker. Good housekeeping means having no unnecessary items in the workplace and keeping all necessary items in their proper place. It includes proper cleaning, disposal of wastes, clean up of spills and maintaining clear aisles, exits and work areas.

HUMAN ERROR
This term is used today to include not just workers' errors, but the human errors with respect to engineering deficiencies and lack of adequate organizational controls. These together account for the majority of incidents.

HUMAN FACTOR
The topic of understanding how people develop behaviours and incorporating features within designs and ongoing operations to prevent errors from being made.

IMMEDIATE CAUSES (also INITIAL CAUSES)
The circumstances that immediately precede an incident or develop during it. Immediate or initial causes, which generally include substandard practices and/or substandard conditions, are usually easy to identify.

INCIDENT
An undesired specific event, or sequence of events that has or could have resulted in harm to people, damage to property, damage to environment or loss to process, or a combination of all.

INCIDENT INVESTIGATION
The process of systematically gathering and analyzing information about an incident. This is done for the purpose of identifying causes and making recommendations to prevent the incident from occurring again.

INCIDENT RECALL
A system to encourage employees to report all incidents, including near miss incidents, in a no fault/blame atmosphere.

INDUSTRIAL SAFETY AND RISK MANAGEMENT
The integrated approach to the management of the continuous reduction of risk to people, environment, assets and production (PEAP) in the industrial setting. Those who benefit from this risk reduction are company personnel, associated contractors and the public at large.

INITIAL CAUSES
See *Immediate Causes.*

INJURY FREQUENCY RATE
(see also FREQUENCY RATE)
A measure of injury frequency:

Frequency rate = $\dfrac{\text{(number of injuries x 200,000 hours)}}{\text{(total exposure hours)}}$

Note: Total exposure hours = the number of persons x the number of hours worked. 200,000 hours represents the total approximate time that 100 persons would work in one year.

INJURY SEVERITY RATE
A measure of incident severity:

Severity rate = $\dfrac{\text{(number of lost work days x 200,000 hours)}}{\text{(total exposure hours)}}$

Note: Total exposure hours = the number of persons x the number of hours worked. 200,000 hours represents the total approximate time that 100 persons would work in one year.

INTEGRATED SAFETY AND RISK MANAGEMENT PROGRAM
The elements are:

Management Leadership, Commitment and Accountability;
Risk Assessment and Management;
Design and Construction;
Process and Facilities Information, Documentation;
Personnel and Training;
Operations and Maintenance;
Management of Change;
Contractors;
Incident Investigation and Analysis;
Community Awareness and Emergency Preparedness;
Program for Continuous Improvement.

Each element is defined and has specific standards and objectives. They form the basis for designing, constructing and operating the company's facilities and for stewardship of performance.

LOSS CONTROL OR LOSS PREVENTION
Measures taken to prevent and reduce loss. Loss may occur through injury and illness, property damage, and poor work quality.

LOSS CONTROL REPORTING (L.C.R.)
A system to report all losses to people, environment, assets and production.

LOST TIME INJURY (L.T.)
Absence from work for more than one workday.

MEDICAL AID INJURY (M.A.)
Injury attended to by medical doctor but of a severity that allows the injured person to return to the job on the same day of injury.

MONITORING
Any activity that is intended to detect deviations and potential risks, ranging from visual checks to hi-tech sensing systems. Monitoring should be the responsibility of staff at all levels, every day.

NEAR-MISS
An incident that could have resulted in a loss, but did not. A near miss is included as an incident. (See **Incident**.)

PERSONAL PROTECTIVE EQUIPMENT (PPE)
Any device worn by a worker to protect against hazards. Some examples are: dust masks, gloves, ear plugs, hard hats and safety goggles.

ppm
Parts per million. A means for expressing low concentrations of pollutants in air, water, soil, human tissue, food or other materials, according to the fraction of mass or volume occupied by the pollutant, i.e., one part salt in one million parts water. Other measures may include ppb (parts per billion) or ppt (parts per trillion).

PRACTICES
The carrying out of the well defined and established procedures.
Note: Putting procedures into practice is an area that needs very meticulous attention.

PREVENTION (ENVIRONMENTAL)
An activity aimed at reducing, to the extent feasible, the release of undesirable substances.

PREVENTATIVE MAINTENANCE
A system for preventing machinery and equipment failure through:

> knowledge of reliability of parts;
> well maintained service records;
> scheduled replacement of parts; and
> maintenance of inventories of the least reliable parts and parts scheduled for replacement.

PROCEDURES
Step-by-step description of how to do a task, job or activity safely and efficiently.

PROCESS
Any activity involving the production, manufacture, use, storage or movement of potentially hazardous materials and/or equipment.

PROCESS CHANGE
Any modification involving substitute materials, other than "in-kind" equipment replacements, or the operation of facilities at conditions outside the established process, mechanical or technical design envelope.

PRODUCTION INTERRUPTIONS
Sporadic or chronic interruptions.

RECOMMEND
To present as worth of acceptance or trial.

RESIDUAL RISK
After a risk is known and accepted and controls are put in place to prevent the incident from happening, including emergency plans to mitigate the incident to a safe level should it happen. It is the risk that remains.

RISK
The possibility of injury, loss or environmental incident created by a hazard. The significance of risk is a function of the probability of an unwanted incident and the severity of its consequence.

RISK ANALYSIS
The use of available information to estimate the risks of a hazard to individuals or populations, property or the environment. Risk analysis, which is used mostly when the potential losses are critical and need to be identified in absolute terms, generally contain the following steps: scope definition, hazard identification and risk estimation. Risk analysis is an objective and formally structured approach to doing a risk assessment. Input is received from several different sources. The analysis tends to be rigorous, methodical and time consuming.

RISK ASSESSMENT
The process of risk analysis and risk evaluation; that is, the quantification and ranking of risks in an objective, informal and user-oriented way. Risk assessment precedes a decision to mitigate or control a risk. Otherwise, all risks would be treated equally.

RISK CONTROL
The process of making decisions about managing risk and implemented, enforcing and re-evaluating the effectiveness of those decisions from time to time. (The results of risk assessment are used in this process.) Risk controls (for example, safety practices, procedures and training) mitigate but never totally eliminate a risk.

RISK EVALUATION
The stage at which values and judgments enter the decision-making process, explicitly or implicitly. A range of alternatives for managing risks is identified by considering the importance of the estimated risks and the social, environmental and economic consequences. Risk evaluation puts the risk into perspective and determines whether the risk is acceptable or needs to be acted upon. This step is influenced by the organization's policy on acceptable risk and the cost/benefit of risk reduction.

RISK IDENTIFICATION
The recognition of factors or conditions which could promote failure or loss. This step is often experience-driven but must also adapt to new inputs. Risk identification is actually the trigger for the risk management process to begin.

RISK MANAGEMENT
The complete process of understanding risk, risk assessment and decision making to ensure effective risk controls are in place and implemented. Risk management begins with actively identifying possible hazards leading to the ongoing management of those risks determined to be acceptable.

RISK UNDERSTANDING
Any public or private communication that discusses the risks.

ROOT CAUSES
See *Basic Causes.*

SEVERITY RATE
See *Injury Severity Rate.*

STEWARDSHIP
The conducting, supervising or managing of the **controls** put in place to reduce risk. The management process needs to ensure the methods used are always done to management standards.

SUSTAINABLE DEVELOPMENT
Development that meets the needs of the present without compromising the ability of future generations to meet their own needs. (UN World Commission on Environment and Development)

SYSTEM (ACTIVITIES)
A set of **steps** or **activities** taken to ensure that stated objectives are achieved. A typical system includes *consideration* of these key elements; agreed *objectives* and documented *procedures*; resources responsible and **accountable** for the implementation and execution; a *measurement* process to determine if desired results are being achieved; and a *feedback* mechanism to provide a basis for further improvement.

SYSTEM (PHYSICAL)
A bounded, physical entity that achieves in its environment a defined objective through interaction of its parts. This definition implies that:

> the system is identifiable;
> the system is made up of interacting parts of subsystems;
> all the parts must be identifiable; and
> the boundary of the system can be defined.

When the word 'system' is used in a physical sense it implies the boundaries of the risk analysis, i.e., what types of hazards are there to be analyzed, from what source and what are the receptors of concern. It is a general term to apply to a wide range of hazards, including both natural and technological (industrial facilities, transportation of dangerous goods, toxic chemicals present in the environment, pharmaceuticals). In terms of technological hazards a system is normally considered to be formed from physical subsystems, their management and the environment.

TASK
A set of related steps that make up a discrete part of a job. Every job is made up of a collection of tasks. For example, answering a phone or entering data into a computer are tasks of an administrator's job.

TASK ANALYSIS
A technique used to identify, evaluate and control health and safety hazards linked to particular tasks. A task analysis systematically breaks tasks down into their basic components. This allows each step of the process to be thoroughly evaluated. Also known as **job hazard analysis.**

TOTAL EXPOSURE HOURS
Total exposure hours = the number of persons x the number of hours worked. 200,000 hours represents the total approximate time that 100 persons would work in one year.

UNDERSTANDING
See *Risk Understanding.*

WebCT
A tool for the creation of web-based learning environments.

Introduction

CHAPTER 1

Industrial Health, Safety and Risk Management
THE INTEGRATED APPROACH

■ INTRODUCTION

The overall concept of Industrial Safety and Risk Management (ISRM) is to provide programs, training and continuous practice in the reduction of risk to people (P), environment (E), assets (A) and production (P) in the industrial setting. (PEAP is used as an acronym for these four areas of concern.) This integrated approach has shaped modern industry and government legislation for safety and risk management practices.

The best industry companies have realized for some time that damage to any or all four areas of PEAP can be very costly to their operations. These companies diligently work at first-class safety and risk management practices since a successful ISRM program can have a major impact on their bottom line. The poor performing companies tend to be forced into at least observing a minimum practice of safety and risk management as dictated by government legislation. ISRM legislation came about by public pressure over the years, particularly after major disasters occurred, especially those with resulting deaths or causing long-term health issues.

■ HISTORICAL BACKGROUND

In the early part of the Industrial Revolution the primary focus was on the machines that were capable of mass production at a high rate, compared to human hands. Most of the machines were designed for maximizing the production and there was very little thought given to the people/machine interface. Not only were the machines not designed to accommodate the worker, but also rarely did he or she receive in depth training in their operation. Furthermore, the working conditions were poor. There were no practices or procedures and no concern from management of the health and welfare of the workers; it was all about production. Workers, adults as well as children, were forced to work long hours. A workday of 11 to 13 hours was normal. Towards the end of the day the chances of mistakes would increase, which could lead to injury and perhaps death of the worker. However, from the management point of view, the workers were mostly expendable and there were many more to replace them.

All did not share the view that workers were expendable. Humane thinkers of the early 1900s, who were connected to government, churches and public authorities, believed it was a real problem. To study the extent of the accident problem in the United States, a survey of the Pittsburgh steel industry in Allegheny County, Pennsylvania, was conducted. It was too difficult to conduct a country-wide study at that time. The results of the survey showed that during a twelve-month period from 1906 to 1907, 526 workers died in industrial accidents. The findings from the Pittsburgh study were made public in the United States. The results of the survey created an awareness among the public who started pushing for changes. The authorities could no longer ignore the problem; it had to be addressed and over the next ten years some very significant and positive changes were made throughout North America.

The first worker compensation law was passed in Wisconsin in 1911. Canada followed soon after with the passing of the Ontario Workmen's Compensation Act of 1915. Safety conscious organizations began to evolve;

the National Safety council was formed in New York in 1913; American Society of Safety Engineers was created in 1911; and the Industrial Accident Prevention Association (IAPA) was formed in Ontario in 1917. The US Bureau of Labor Statistics was formed to tabulate and publish facts on industrial accidents and the US National Bureau of Standards began to publish safety standards for materials and equipment.

At this point in time "safety" meant the absence of injury. The solutions and/or measures to prevent injuries were most often of a hardware character, such as machine guarding, warning signs, hard hats and safety glasses. Very little preventative work was done involving the people interfacing with the machines. In the early 1930s Wilfred Heinrich, a European scientist, came up with a different approach to industrial accident prevention. Heinrich's concepts included the theory that the methods of most value in accident prevention are analogous with the methods for control of quality, costs and production. In addition, he indicated that management has the best opportunity and ability to initiate the work of prevention and therefore should assume the responsibility. Heinrich suggested an integrated approach to the prevention of accidents. Heinrich's philosophy was only followed to a limited degree and unfortunately the emphasis was placed on some of his other concepts that said the unsafe acts of persons were responsible for a majority of accidents. This gave management the opportunity to blame the worker.

The approach to safety as the "freedom of injury" changed over the years and in the 1940s to 1960s, the health and safety of the worker was incorporated in the overall management system with the emphasis on accident prevention and control. Safety took on a new meaning: "Safety—the control of accidental loss". This concept of total loss control involved the examination of all loss exposures in a company in order to obtain maximum benefits and the development of practices and procedures to reduce all losses involving people, environment, assets and production. Furthermore it was realized that incidents happen because of management systems breakdown. This in turn removed much of the blame but not the responsibility from the worker.

The industries that have adopted this integrated approach have seen a reduction in incidents and losses and in addition improvements in both production and quality and reliability. A significant outcome of this was an increase in quality of life in the work place among all employees at all levels. The best of industry companies implements and demands this approach of all employees at all levels. As an engineer, geologist or geophysicist it is important to understand the philosophy and obtain the knowledge behind the integrated approach to industrial safety and risk management in order to play an active role in minimizing the losses to people, environment, assets and production. This concept of total loss control involved the examination of all loss exposures in a company. This was done to obtain maximum benefit and the development of practices and procedures to reduce all losses involving people, environment, assets and production (PEAP).

■ THE INTEGRATED APPROACH AND BENEFITS—PEAP

Industrial Safety and Risk Management is defined as the integrated approach to the management of the continuous reduction of risks to people, environment assets and production in the industrial setting. Those who benefit from this risk reduction are company personnel, associated contractors, the public at large and investors. Figure 1–1 provides a definition for industrial safety and risk management.

There are several benefits to an integrated industrial safety and risk management program. For the most important element, **People**, there is the elimination of health problems, injuries and death, which will alleviate the short-and long-term suffering to victims and families. This primarily applies to people in the workplace including company employees, contractors and visitors. But it also applies to any member of the general public who has the possibility of being affected by an incident.

Another benefit is the protection of experienced and well-trained valuable employees. An efficient safety and risk management program will also attract the best employees and keep turnover rates low. Training and retraining will ensure that the employees are competent, which will reduce the risk of injury and health problems. There is also a positive effect on morale among employees, company climate, greater job satisfaction and reduced absenteeism.

The second element is the protection of the **Environment**, which means protection of water, air and land. Protection of the environment encompasses activities within the fenceline of the operation, any possible impact on public or neighbouring property and transportation corridors used to receive raw materials and send out product. This does not only include losses during upset and accidental releases, but also during normal day-to-day operation.

The benefits to **Assets**, the third element, are: A reduction in damages to equipment, plant and transportation systems and can also include reduction in losses to feed stocks, products and other company materials, etc. As a function of both these and the efforts in the areas of the other elements, there will be a positive effect on profits and stock value.

The benefits to the last element, **Production**, is the reduction in lost production due to delays and interruptions resulting in increased productivity.

Supply is reliable and stable, which most often results in a strong position in market share. Overall an excellent safety performance most often indicates a well run efficient and effective company. It also provides a positive influence on the plant, company and industry image.

■ COMPANY POLICY STATEMENT AND POLICIES

A company policy statement is very important for the success of a safety and risk management program. The policy should clearly state responsibility and accountability expectations. It should state that management is committed to the program and it should give both employee and contractor direction. A policy statement also plays an important role in increasing the public's understanding of the overall goal of the company. It should be kept current at all times and be available on prominent display all through the company work areas. Figure 1–2 shows a typical company policy statement.

■ VALUE SYSTEM AND COST/BENEFIT

It is impossible to have zero risk in any situation. There has to be a balance between the resources spent on achieving low risk versus the benefits. If the costs of the risk reduction are higher than the benefits, then the risk reduction study should be rethought. The acceptable level of risk depends on the type and size of the business or company. Each company must decide standards, acceptable levels of risk and capacity to absorb losses without major effect to the bottom line.

Losses regarding people should be the same for all companies, that is losses

FIGURE 1–2: AN EXAMPLE OF A POLICY STATEMENT

"ENERGY PRODUCTS" LTD. POLICY STATEMENT ON INDUSTRIAL SAFETY AND RISK MANAGEMENT

Industrial Safety and Risk Management is an integrated and consistent approach toward the elimination of incidents and the reduction of risks to people, production, facilities and the environment. These activities must be focused on our Company, the contractors and the associated public-at-large.

The Company is fully committed, on a continuous basis, to the application of our programs on Industrial Safety and Risk Management in the total operation of our business.

The Company will provide safety and health working conditions while demonstrating excellence in incident, fire and security protection and compliance with laws, regulations and procedures.

All employees, contractors and visiting public must comply with the Company's rules, regulations and procedures.

The Company recognizes that excellence in Process Safety and Risk Management can only be achieved through the active participation of everyone at all levels, including contractors. All of this must be fully integrated with everyday activities and not treated as a separate issue.

Fiona McTavish
President and Chief Executive Officer
January, 2001

in this area are not acceptable and everything should be done to reduce the risks to very low levels. The same should be said for the environment as well but, for the time being, it is acceptable for companies to emit certain substances in certain amounts to the environment where the standards and limits are set by Environment Canada and provincially, for example, by Alberta Environment. Emissions of very potent pollutants should be stopped at all times and the risks should be reduced to very low levels. Regarding assets and production losses, this has a very profound effect on the profit margin of a company and it is in the company's own interest to reduce the risk within these areas.

Overall, the acceptable economical losses are mostly decided by the company, whereas losses involving people and the environment are not acceptable in today's society. It is no longer viewed

as being part of doing business; they are losses that can be avoided.

■ SUMMARY

In this chapter, we have strongly emphasized the advantages of the integrated approach to safety and risk management showing the benefits to be gained. Although an awareness was growing about safety issues in the early 1900s, it took until the 1970s before the major first-class companies commenced practicing this approach. Then many mid-size to smaller companies followed by example. However, there are still a number of companies of all sizes who do not follow this integrated approach or neglect using it. These companies invariably get into trouble causing damage to people, environment, assets and production—PEAP.

Over the years from the 1900s to the early 1970s, the best of industry companies and government bodies built up a new language for the field of industrial safety and risk management. This consisted of creating a new vocabulary of terms and definitions so that they could easily communicate with each other. It also allowed improvement in the systems to be developed much more effectively and quicker. This professionalism in the field also helps greatly in providing a basis for sound legislation and professional considerations and benefits (e.g., engineers, business managers, medical experts).

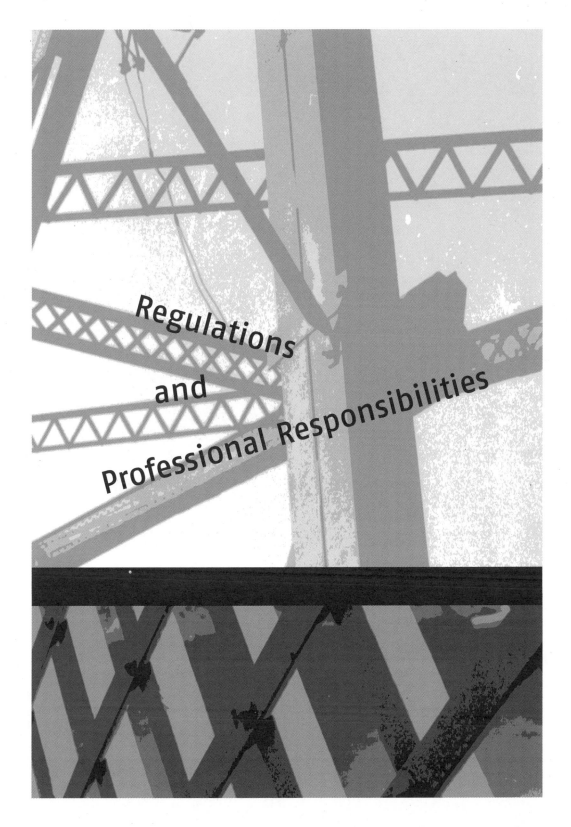

Regulations
and
Professional Responsibilities

CHAPTER 2

Government Regulations

■ FEDERAL AND PROVINCIAL LEGISLATION

Government acts and the corresponding regulations represent the public's requirements for companies to conduct business. These regulations can be changed and influenced by many sources. The influence can be specific to the local community or global—particularly when it comes to environment or public safety—in nature. Because of this global approach, countries seemingly are developing their laws and associated regulations in line with one another. Countries are learning from each other and adopting similar rules for their public. The result is a need for a company that operates in several different jurisdictions to reflect these global needs in their policies in order to be successful.

In Canada the responsibility for these regulations lies within all three levels of government—federal, provincial and municipal. Generally most of the responsibility lies with the provinces with the federal requirements specific to specific acts the federal government administers (see appendix in Chapter 16 for a detailed listing). The municipal (or local) jurisdictions have their own requirements spelled out in bylaws they develop, which are mostly focused on the residents of the community and protection of their quality of life.

Provincial legislation and regulations are handled differently but the fundamental issues these regulate are similar if not the same. Chapter 16, "Provincial Regulatory Agencies and Workers' Compensation Boards" goes into more detail. For a company working in several provinces their policies can be developed to work effectively. This is a challenge for any company but as long as the policies reflect the company's activities in the fundamental areas the company will be effective in managing safety and risk in the workplace. Regulations are written and administered differently in each province but all address the fundamental topics.

All provinces in Canada have different health and safety systems and legislation. These produce similar results, ensuring workplace health and safety, but there are very different regulations province to province. If you are working in another province other than your home province, you must connect with the appropriate Government agency for health, safety and environmental protection. They will have regulations, publications and standards that you must follow in any work that you undertake in that province.

■ OTHER COUNTRIES SYSTEMS AND LEGISLATION

There are different regulations for all the states in the United States and other foreign countries. Some countries, especially in the third world, are not as strict in their standards and required responsibilities. As a professional, you have to meet the standards of the state, province or country that you are working in. We believe it is advisable to work to the highest standard that you are accustomed to. That is, if the country you are working in has a lower standard than that of Canada, use the Canadian systems and legislation.

■ ALBERTA LEGISLATION

This section summarizes the most important points of provincial legislation from a safety and risk management perspective. Alberta Human Resources and Employment provides an excellent model of legislation that can be studied.

Occupational Health and Safety Act of the Province of Alberta

The Act as a whole is very important and must be read, understood and complied with by managers, supervisors and workers. There are three sections that are highlighted and paraphrased or quoted in this section:

> obligation of employers, workers, etc. (section 2);
> prime contractor responsibility (section 3); and
> existence of imminent danger (section 35).

Note: The printed act published by Queen's Printer Bookstore is the legal document to be followed and the most recent version is required. A copy of the *Occupational Health and Safety Act of the Province of Alberta* (December 2002) can be found at http://www.qp.gov.ab.ca/documents/Acts/002.cfm?frm_isbn=0779714946

Section 2: Obligations of Employers and Workers

Section 2 of the Act states that every employer must take any reasonable practical step to ensure the health and safety of the employees and any other persons present on the work site. The employer shall also ensure that the employees are, according to section 2(1)(b), "aware of their responsibilities and duties under this Act, the regulations and the adopted code."

According to section 2(2)(b), every worker is obligated to "take reasonable care" to protect his or her health and the safety of other workers present during the work. Furthermore, every worker shall work together with the employer to ensure the health and safety of himself/herself and other workers on the site, including workers employed by a different employer (contractors, etc.).

Section 2(3) states that "Every supplier shall ensure, as far as it is reasonably practicable for the supplier to do so, that any tool, appliance or equipment that the supplier supplies is in a safe operating condition." As well, the "supplier shall ensure that any tool, appliance, equipment, designated substance or hazardous material complies with this Act, the regulations or the adopted code" according to section 2(4).

Section 3: Prime Contractor

According to section 3(2) the "prime contractor for a work site is (a) the contractor, employer or other person who enters into an agreement with the owner of the work site to be the prime contractor or, (b) if there is not agreement, the owner of the work site." Section 3(1) states a prime contractor is required "if there are two or more employers involved at work site at the same time." It is the responsibility of the prime contractor, according to section 3(3), to "ensure, as far as it is reasonably practicable to do so, that this Act and regulations are complied with in respect of the work site." Section 3(4) states that the prime contractor shall "do everything reasonably practicable to establish and maintain a system or process that will ensure compliance with this Act and the regulations in respect of the work site."

Section 35: Existence of Imminent Danger

This section is very important and should be studied in its entirety, but only the first part is quoted here.

35 (1) No worker shall
   (a) carry out any work if, on reasonable and probable grounds, the worker believes that there exists an imminent danger to the health or safety of that worker,
   (b) carry out any work if, on reasonable and probable grounds, the worker believes that it will cause to exist an imminent danger to the health or safety of that worker or another worker present at the work site, or
   (c) operate any tool, appliance or equipment if, on reasonable and probable grounds, the worker believes that it will cause to exist an imminent danger to the health or safety of that worker or another worker present at the work site.

(2) In this section, "imminent danger" means in relation to any occupation
   (a) a danger that is not normal for that occupation, or
   (b) a danger under which a person engaged in that occupation would not normally carry out the person's work.

Other very important sections in the *Occupational Health and Safety Act* are listed below. These are sections that will require active participation from engineers, geologists and geophysicists:

> Multiple obligations (section 4);
> Serious injuries and accidents (section 18);
> Regular inspection of work sites (section 25);
> Report on designated substances (section 29);
> Controlled product (section 30);
> Joint work site health and safety committees (section 31);
> Written health and safety policies (section 32);
> Code of practice (section 33);
> Where disciplinary action prohibited (section 36); and
> Offences (section 41).

It must be emphasized that engineers, geologists and geophysicists must read, understand and comply with the entire *Occupational Health and Safety Act of Alberta* at all times. A copy of the Act can be obtained from the Queen's Printer or from the Alberta Government's Web site.

Environmental Protection and Enhancement Act

It is a requirement to know and comply with this Act in its full length and the engineer, geologist and geophysicist should also know the specific regulations that are related to his/her working area. See also Chapter 3 "Due Diligence" and the "Environmental Practice, a Guideline" published by APEGGA as a supplement to the Professional Code of Ethics.

Due Diligence

During the 1980s and early 1990s, environmental and workplace health and safety regulatory requirements significantly increased in Canada. These changes resulted in the maximum fines for infractions being substantially increased, some as high as one million dollars. In addition to fines, individuals can also be imprisoned for infractions on environmental and workplace health and safety legislation.

Due to this increased liability, many Canadian corporations have changed their business practices in order to minimize their risks of fines and imprisonment of key personnel. These corporations have found that their best defense (minimizing their risk of fines and imprisonment) is **due diligence.**

■ DEFINITION OF DUE DILIGENCE

In 1978, the Supreme Court of Canada created the defense of due diligence in a decision involving the City of Sault Ste. Marie. Prior to the Supreme Court of Canada's decision on the Sault Ste. Marie case, health, safety and environment statutes were regarded as absolute liability statutes. This meant that if a health, safety or environmental incident occurred, the defendant (whether it was a corporation or an individual) was guilty no matter what had been done to prevent it from occurring.

As a result of the Supreme Court of Canada's decision in the Sault Ste. Marie case, the defendant (corporation or individual) charged with an offense is permitted a defense of due diligence. To show due diligence the defendant must establish that all reasonable steps were taken to prevent the infraction.

Many courts across Canada have said due diligence, or all reasonable care, involves considering the steps a reasonable person could have taken in the circumstances:

1. Was there an effective risk management system (to prevent ill health, pollution and workplace injuries) in place prior to the offense?
2. Was the system operating effectively?
3. Was the system being maintained?
4. Did the accused person reasonably but mistakenly believe in a set of information which, if true, would render the actions or lack of actions innocent?

Canadian courts apply a number of criteria when determining the standards of care required demonstrating due diligence. These include:

› Industry standards common to the work being done.
› Special standards that might dictate a higher level of care required:

a) The degree of knowledge or skill expected of the person.
b) The location of the operation (an example is a highly sensitive environment).
c) The severity of potential harm.
d) The extent to which the underlying causes of the offense were beyond the control of the accused, and

e) The alternatives, i.e., what was done against what could have been done.

Achieving Due Diligence

The key of meeting the test of due diligence is the development, implementation and maintenance of an effective health, safety and environmental management program. These can be somewhat different for the different industries and companies. However, Chapter 5, Industrial Health, Safety and Risk Management Programs provides a good example of a proven program.

Key Lessons to Be Learned

Due diligence is more than just a legal issue. It is a standard that corporations and individuals can use to assess the content and effectiveness of their safety, health and environment management program. Most successful corporations believe that the major benefit of applying due diligence to workplace health, safety and environment is not the legal defense it provides but the many other benefits that an effective safety, health and environment management program provides the corporation, its workers and the community.

Why Care About Due Diligence?

Due diligence is important as a legal defense for a person charged under occupational health and safety legislation. If charged, a defendant may be found not guilty if he or she can prove that all precautions, reasonable under the circumstances, to satisfy his obligations, were taken to protect the health and safety of all workers. This is known as the "defense of due diligence".

What Is Meant By Due Diligence?

Due diligence is the level of judgment, care, prudence, determination and activity that a person would reasonably be expected to do under particular circumstances. Applied to occupational health and safety, due diligence means that employers shall take all reasonable precautions, under the particular circumstances, to prevent injury or accidents in the workplace. To exercise due diligence, an employer must implement a plan to identify possible workplace hazards and carry out the appropriate corrective action to prevent accidents or injuries arising from these hazards.

Liabilities

Strict Liability—Express possibility of right to choose to do something. You have the option of deciding if you do or do not proceed based upon the circumstances.

Absolute Liability—Obligation or command to do something. You do not have a choice, you must do it.

Reasonably Practical

This general statutory obligation conveys a message to the courts that the standard of care within occupational health and safety legislation is not an absolute obligation, but a strict obligation. While it is not practicable to take precautions against a danger which is not known to exist, once a danger is known it should become reasonably practicable to do something about it. In the eyes of the law, if something is reasonably practicable then it must be done. It is up to the employer to find a reasonable and practicable means to reach the objective and to continue to keep up-to-date on new developments in the field.

What Is the Standard of Due Diligence?

The standard of care required to comply with the occupational health and safety regulations is taking all reasonable precautions in the circumstances to protect the well-being of workers:

> provincial and national standards;
> industry practices; and
> manufacturers specifications.

In other words: *If it can be done, it must be done with the technology of today.*

The Law

Every employer shall ensure, as far as it is reasonably practicable for him or her to do so:

a) The health and safety of

  i) workers engaged in the work of that employer, and
  ii) those workers not engaged in the work of that employer but present at the worksite at which that work is being carried out, and

b) that the workers engaged in the work of that employer are aware of their responsibilities and duties underspecific legislation and the regulations.

Penalties

There are penalties for noncompliance that may result in fines or imprisonment:

> first offense—up to $150,000 and/or 6 months in jail;
> subsequent offense—up to $300,000 and/or 1 year in jail.

The human and economic costs can be **far greater** in the event of an accident.

Employer's Responsibilities

There are specific employer's responsibilities to establish due diligence, including:

> Understanding the responsibilities as defined in the occupation health and safety legislation.
> Implementing a system to identify health and safety hazards and take every precaution reasonable in the circumstances to protect the employees.
> Ensuring adequate training of employees by providing information, instruction and supervision to protect their health and safety.
> Appointing a competent supervisor.
> Enforcing health and safety procedures.
> Seeing that equipment, materials and protective devices provided are maintained in good condition and used as prescribed.
> Taking action immediately when you are informed about a potentially hazardous situation (*willful blindness*).
> Initiating immediate investigation into accidents.
> Developing and following emergency response planning.
> Conducting inspections to identify new hazards and monitor the effectiveness of hazard control measures.
> Maintaining written records of all health and safety activities.

Supervisor Responsibilities

There are responsibilities of supervisors in establishing due diligence, including:

> Instructing new workers in safe work procedures.
> Training workers for all tasks assigned to them, and checking their progress.
> Ensuring that equipment and materials are properly handled, stored and maintained.
> Ensuring that only authorized, adequately trained workers operate tools and equipment and use hazardous chemicals.
> Enforcing health and safety regulations.
> Correcting unsafe acts.
> Acquainting the workers with any hazards in the workplace.
> Formulating safe work procedures.
> Frequently inspecting for hazards in your area.

Failures to Follow Due Diligence

Common failures to follow due diligence are:

> Failing to train workers properly.
> Equipment is not properly maintained.
> Allowing the wrong person to do the job.
> Having the wrong equipment for the job.
> Having insufficient people on the job.
> Not knowing that a danger exists.
> Allowing workers to be "in a hurry".

■ SUMMARY

Due diligence is the level of judgement, prudence, determination and activity that a person would reasonably be expected to do with respect to managing PEAP in the workplace. Proper policy and procedures and documentation systems need to be in place. In order to show "due diligence" proof of activity must be shown.

Proving due diligence means proving your innocence. This is one of the few legal situations where proving innocence is needed. Normally a court needs to prove guilt.

As professionals we have a responsibility to perform as best as we can. Sometimes the path is a clear one leaving us to our best judgement. Properly evaluating the issue and documenting our choices means a Due Diligence defense can be used.

CHAPTER 4

Professional Organizations and Professional Responsibilities

■ PROFESSIONAL ORGANIZATIONS

According to the *Oxford Dictionary*, a profession is defined as "A vocation or calling that involves some branch of advanced learning or science." A profession includes:

> specialized knowledge;
> intensive preparation;
> high standards of achievement;
> high standards of practice;
> ethical conduct;
> continued study; and
> public service.

In fact to be a professional requires the person to recognize the responsibilities inherent to these points with respect to safety and risk management. Those who are not professionals, especially the general public, depend on those who are for the proper and most effective advice, operation or design. Maintaining high standards and ethical requirements are crucial for a professional.

In order to assist with ensuring the public is looked after, professional organizations provide rules of conduct in terms of codes of ethics. These codes are designed to ensure the requirements of the public are met through the setting of high standards, provision of guidance, maintaining a sense of high standards, clear of conflict and ethical in everyone's eyes.

Sometimes provinces delegate the authority to manage the appropriate legislation for a specific profession, like Engineering, to a professional organization. A professional should always have the public's needs in mind; the direction of the profession should be towards meeting and exceeding those needs and working towards improving on those needs.

Safety and risk management is one of the areas professionals must have as a priority as they perform their duties. The Engineering profession clearly has defined these needs and as the example below shows, actually benefits from doing so. By being a body involved in improving safety and risk management all of society, including the profession, benefit.

■ BENEFITS TO PROFESSIONALS, INDUSTRY AND THE PUBLIC

There are many specific benefits for professionals who are working in an environment with a successful industrial safety and risk management program. The Association of Professional Engineers, Geologists and Geophysicists of Alberta (APEGGA) standards provides a good example of a professional code of conduct and outlines some basic guidelines to follow to achieve a successful and productive professional career.

■ CODE OF ETHICS

Industrial safety is one of the primary focusses of professional rules of conduct. The first rule of conduct in the APEGGA Code of Ethics states:

> Professional engineers, geologists and geophysicists shall have proper regard in all their work for the safety and welfare of all persons and for the physical environment affected by their work.

All qualified APEGGA members or other professionals from other provinces or countries must understand and in their day-to-day work practice all of the basics of industrial safety and risk management. Because of the public's much increased awareness and demands,

it is very important that the professional members are at the forefront of activities to reduce the risks to people, environment, assets and production (PEAP) in an **integrated** manner.

■ RESPONSIBILITY UNDER LAW

The obligations of the employer and the worker regarding the health and safety of the employees and other persons present on the work site are specifically legislated province by province. The key points of the *Occupational Health and Safety Act of the Province of Alberta* (December 2002), discussed in Chapter 2, examine these responsibilities. A professional can be in either category and is certainly seen in this way by the law. This makes it very important that they understand their obligations under the Act as part of being a licensed professional. Both the public and the courts, most often, expect a much higher degree of responsibility requirement from professionals than most other industrial employees.

In addition to this responsibility there are also penalties to be paid for noncompliance to provincial legislation. For example, if offences are committed by responsible people and are proven, the fine for a first offence in Alberta can be up to $150,000 and a further $10,000 per day for the period the offence is continued. This first offence can also result in imprisonment not exceeding 6 months. The maximum fine is $300,000 and a further fine of $20,000 per day for the period the offence is continued and a prison term not exceeding 12 months (see section 32(1) of Alberta's *Occupational Health and Safety Act*).

There are some specific activities regulated federally that can impact safety and risk management (see Chapter 16). Provincial regulations are the most important to recognize but ensuring compliance to appropriate federal regulations cannot be overlooked.

■ REDUCTION OF LOSSES AND HUMAN SUFFERING

Professional engineers, geologists and geophysicists of all disciplines must play a significant role in the continuous improvement of workplace health, safety and risk control (safety and risk management). Their basic education and the scope of their professional careers prepares them for a significant involvement in this whole field as part of their everyday responsibilities and activities.

Industry leaders have become increasingly aware of the need for major improvement in this field. This awareness stems from the concern of the negative effects that have and can occur in operations worldwide. Every year industry suffers enormous and unacceptable losses in the areas of employee health, safety, assets, production and environment. The subsequent effect on company employees, contractors and the public at large, is a significant driving force to provide continuous improvement of performance. The professional engineer, geologist or geophysicist can provide a key influence on this required area for improvement.

For example, in Alberta the **yearly average** losses in industry are approximately as follows:

> industrial deaths = 120;
> serious injuries = 55,000;
> Workers' Compensation Board costs = $800 million; and

> total loss costs = $5 billion (including people, environment, assets and production).

For all of Canada these yearly results are almost 10 times those of Alberta's. Internationally, comparable negative results are occurring in the industrial world. With the current competitive climate (the need to do more with less), it is time for a significant improvement to be made in this industrial field.

■ BENEFITS TO THE PROFESSIONAL AND THE COMPANY

It is important to understand the effect of safety and risk management to the bottom line of company results. If these results are very positive, then benefits flow to both the company and the professional.

By understanding the key role that safety and risk management performance plays in the "health" of a company or organization, the professional will have a much better basis for a successful career in his or her chosen area of expertise.

Being involved in avoiding human suffering to both potential victims and dependents gives the professional major job satisfaction. In addition, the reduction of risk to the environment, assets and production will contribute to a positive professional attitude.

Almost all companies have a performance appraisal system for their employees. In the best of industry companies this system will have approximately 30% of its content based on the employee's performance in safety and risk management. Knowledge and expertise in this whole field is a very important asset in building a first-class career.

FIGURE 4–1: THE SAFETY CYCLE

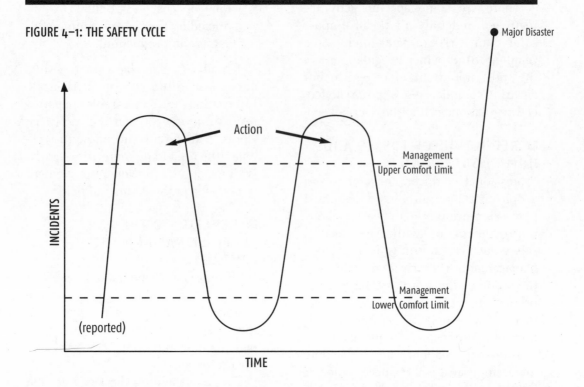

In order to achieve these benefits, there are many key factors to be considered. We believe the following three points are very important.

Continuous Effort

Application of the basics of safety and risk management, both on a day-to-day and on a long-term basis, must be continuous and integrated with your professional work. Usually both individuals and companies expend a lot of energy on safety and risk management activities directly *after* a major incident or series of incident. When their focus is diminished by time, this effort is reduced drastically, until the next major event happens. This produces a sine-wave type of performance, as illustrated in Figure 4–1. This totally negates the continuous improvement and quality management philosophy and practices that most companies and employees strive towards. It certainly reduces the confidence of the workforce in management leadership and commitment. The continuous attention to this whole topic will certainly promote excellence in operational performance and reliability.

Ensuring the Principles Are Practiced

Many companies and individuals do an excellent job at providing procedures, manuals, equipment, training and programs. However, they do not put nearly as much effort into ensuring that all these items are practiced thoroughly and continuously in the field. It is extremely important that these practices are car-

FIGURE 4-2: EFFECTIVE APPLICATION FOR FIRST-CLASS PERFORMANCE

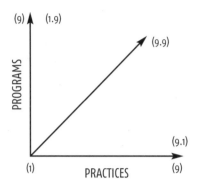

Note: The ideal performance is a (9.9), where there is a high concern for both practices and program. A (1.9) performance indicates high concern for the program, but not for the practices. An example of this is a program that is sitting on a shelf without being applied. High concern for practices, but not for the program is indicated by a (9.1) position in the figure. An example would be where workers carry out non-standardized practices instead of using a written program.

ried out voluntarily, particularly at the times when there is no supervisor or professional person present. That is, a climate for voluntary compliance to practices (how effectively it gets done) must be created and sustained by the management and the professionals. Without this the very best programs, procedures and manuals will not advert disaster (see Figure 4-2).

Professional Reputation

It is very important for professionals to build and protect their professional reputation. The APEGGA Code of Ethics covers this area specifically. Under the topic of industrial safety and risk management it is extremely important to protect your own health and ethical reputation. The connection between your reputation and your responsibilities within the industrial safety and risk management area is clearly defined in the APEGGA Code of Ethics earlier in this chapter and the Alberta *Occupational Health and Safety Act* (Chapter 2).

In our professional experience the major blunder that companies make with programs is that their design is usually good but their application is poor. That is there is not enough effort on a continuous basis, particularly from senior management, to ensure that the program is being used, employees are being trained and it is reviewed for continuous improvement.

Industrial Safety

and

Risk Management Programs

Industrial Health, Safety and Risk Management Programs

■ INTRODUCTION

Most industrial companies, of all sizes, find it necessary to have an industrial safety and risk management program in order to achieve first-class results. Companies can obtain these programs from various sources, i.e., self developed with high involvement from employees, purchased, etc. However, the particular program they use must be customized to suit the company's objectives and style of operation. It also must fit the size of the company, location and type of business. All of this is required so the whole organization takes ownership of a responsibility for the program. That is, all employees should use it as a framework for carrying out their individual work assignments.

All of these types of programs contain a number of key elements of areas of responsibilities in safety and risk management. The number of elements can range from approximately six to twenty depending on size of the company, type of operation and length of time from initial implementation. Usually companies start off with a few key elements and expand with additional elements as the program matures.

Each program element must be well defined and have specific standards and objectives. The overall program should form the basis for designing, constructing and operating the company's facilities. It should also form the basis for stewardship of performance at the company, department and individual employee level.

■ A TYPICAL INDUSTRIAL HEALTH, SAFETY AND RISK MANAGEMENT PROGRAM

The program discussed below has been developed by studying the programs of best of industry companies and with reference to the program designed by the International Loss Control Institute (ILCI) (now called Det Norske Veritas, Incorporated (DNV)). Other programs that are being used with success are CAER (Community Awareness and Emergency Response), API (American Petroleum Industries), OSHA 1910 (Occupational Safety and Health Administration), and ISO 9000 (The International Organization of Standardization). These programs can be researched at provincial government departments, libraries or on the Internet.

The particular program described here consists of 11 elements that have been proven to be effective in avoiding losses and reducing risks to people, environment, assets and production. The list of elements is presented below, following by a short description of each.

Elements*

ELEMENT 1: MANAGEMENT LEADERSHIP, COMMITMENT AND ACCOUNTABILITY

ELEMENT 2: RISK ASSESSMENT, ANALYSIS AND MANAGEMENT

ELEMENT 3: DESIGN, CONSTRUCTION AND START-UP

ELEMENT 4: OPERATIONS AND MAINTENANCE

ELEMENT 5: EMPLOYEE COMPETENCY AND TRAINING

ELEMENT 6: CONTRACTOR COMPETENCY AND INTEGRATION

ELEMENT 7: MANAGEMENT OF CHANGE

ELEMENT 8: INCIDENT REPORTING, INVESTIGATION, ANALYSIS AND ACTIONS

ELEMENT 9: OPERATION AND FACILITIES INFORMATION AND DOCUMENTATION

ELEMENT 10: COMMUNITY AWARENESS AND EMERGENCY PREPAREDNESS,

ELEMENT 11: PROGRAM EVALUATION AND CONTINUOUS IMPROVEMENT

ELEMENT 1: MANAGEMENT LEADERSHIP, COMMITMENT AND ACCOUNTABILITY

This is the most important element of any program and without it the program will fail or at best produce mediocre results. It is management that provides the perspective, sets the goals and allocates the resources to ensure a successful and reliable operation. In addition, the management plays a key role in planning, organizing, leading and stewarding the safety and risk management program. This program should be integrated with all other company activities, i.e., part of the business plan. It should receive short-term and long-term planning and attention.

Management will come to this commitment position by realizing that people are their most important asset. Also, by understanding that safety and risk management provides a significant opportunity for managing costs and improving operational reliability.

Some examples of how the objective of this element can be met are as follows:

Leadership by example: All members of management must practice all the safety and risk management directives, procedures and rules,

---

* This is not the ILCI Program, which has approximately 20 elements. If required refer to DNV.

etc., in their daily work in order to show all employees their strong commitment.

Management visibility in the field: It is very important that management visit the work site on a planned basis during normal operation and not just when things go wrong. They should interface with the workforce and actively listen to any key issues that they may have. These issues should be resolved as soon as possible. All of this can be done at toolbox talks, site inspections, monthly safety meetings, etc.

Objective settings and stewardship: Management should continuously provide leadership in setting objectives and ensure that the results are stewarded against these objectives. Where deviations exist, action items should be applied as appropriate. This process should have the involvement of the whole organization and be very visible.

"Line"* responsibility: Management should constantly emphasize that safety and risk management is a *"line"* responsibility. Safety professionals should be used as a resource but they are not in charge of safety. Every employee must be encouraged to be his or her own "safety officer".

Management participation: All members of management should drop in on various safety and risk management activities taking

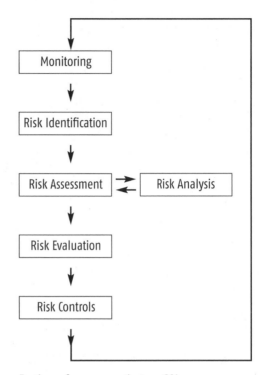

FIGURE 5–1: RISK MANAGEMENT FLOW CHART

Further references are the two GSA documents "Risk Analysis Guidelines" and "Risk Management Guidelines".

place in different parts of the organization. For the most part this should be done as a participant, not always as a leader.

Management's key objective is to create a safety and risk management culture that will involve the whole organization in personal commitment to take on personal responsibility and accountability. This will ensure safe and reliable operation, including continuous improvement.

---

*Line = Front-line employees

ELEMENT 2: RISK ASSESSMENT, ANALYSIS AND MANAGEMENT

Risk management is the process for elimination of hazards and reduction of risks to people, environment, assets and production in any setting. It is complementary to the risk management process, except that it focuses exclusively on losses that have not yet occurred (proactive approach). It can address such areas as physical facilities, procedures, work practices, human factors and organizational changes, etc.

The risk management framework depicts the essential steps in the whole risk management process. It includes the following steps: system monitoring, risk identification, risk assessment, risk analysis, risk evaluation and risk control (see Figure 5-1 Risk Management Flow Chart). This particular topic is discussed in depth in Chapter 7.

ELEMENT 3: DESIGN, CONSTRUCTION AND START-UP

These activities have historically been shown to be of high risk for the most part because safety and risk management was not integrated with the standards, procedures and practices throughout the whole process of design, construction and start-up.

The appropriate management teams must ensure that the essential safety and risk management objectives are built into their procedures, practices, standards, systems, etc. The results should be monitored and stewarded with the aim of continuous improvement. This can be markedly improved by the involvement of all their personnel.

The specifics of this element must be customized for the particular company and project. However, it should contain safety and risk management criteria and objectives in such areas as: design standards and practices; project management; risk assessment; quality control; pre and post start-up reviews and activities.

ELEMENT 4: OPERATIONS AND MAINTENANCE

In order to control risk it is essential to operate and maintain facilities within established criteria. This requires effective procedures and thoroughly established practices and qualified personnel who consistently execute these procedures and practices. It also requires structured inspections and maintenance systems, reliable safety systems and control devices.

In addition, this element should include timely and accurate implementation of a system that ensures the update of any changes made in operations and maintenance, standards, practices, procedures and designs, etc. It should also include incorporation of checks and authorizations through a work permit system and implementation of special procedures for managing higher risks operations.

A system should be in place to track hazardous emissions and wastes in order to comply with environmental objectives and regulations. Special attention must also be given to the efficient and effective abandonment of facilities according to environmental regulations.

A key objective of this element is to ensure safe and reliable operation.

ELEMENT 5: EMPLOYEE COMPETENCY AND TRAINING

The successful management of operations depends on people. Operations that are safe and environmentally sound require careful *selection, placement,* ongoing *assessment* and effective *training* of employees. Management must make certain that systems are in place and effectively used to cover this important element. This includes ensuring sufficient budget and resources are made available through all types of economic climates. Special attention must be given to make sure that safety and risk management criteria are built into this whole process, otherwise the operation cannot be successful.

ELEMENT 6: CONTRACTOR COMPETENCY AND INTEGRATION

Contractors who carry out work on behalf of the owner company can have a major impact on the company's operations and its reputation. It is absolutely essential that they perform their work in a manner that is consistent, compatible and integrated with the owner company's standards, policies, procedures, practices and business objectives.

First-class performance in this area is obtained by having *systems for evaluation* and *selection of contractors* that includes an *assessment of their capabilities* to perform work in a safe and environmentally sound manner. It is essential that management define, communicate and include a system for self-monitoring and stewardship of results. Feedback is essential to ensure deficiencies in performance and quality of work are corrected on a continuous basis.

The fact that the performance of a company is only as good as its weakest link may seem to be common sense. In practice this weak link can be the working relationships between the company and the contractors. Contractor personnel, if selected correctly, normally come

with excellent skills in their particular field. However, you cannot expect them to have the same knowledge and understanding of the company's facilities and systems as the operations personnel. This is where the high risk can occur. In the last 20 years a number of serious disasters have occurred in industrial settings around the world with the major contributing cause being deficiencies in owner company/contractor interface and relationships. See Chapters 10 and 11 for more detail.

ELEMENT 7: CHANGE MANAGEMENT

Changes in operations, procedures, site standards, environment, facilities, personnel, organization, etc., must be evaluated and managed to ensure that safety and environmental risks arising from these changes remain at an acceptable level. Similarly, changes in laws and regulations must be reflected in facilities and operating practices to ensure ongoing compliance. Management must ensure that systems are in place for both temporary and permanent changes. Procedures for managing change address the following:

> identification and assessment of changes;
> authority for approval of changes;
> acquisition of needed permits;
> communication of potential consequences and required compensating measures;
> analysis of safety and environmental implications
> documentation, including reasons for change;
> time limitations; and
> training.

Major changes are normally well managed particularly if a project manager is in place to provide coordination and leadership. However, the very numerous small changes that are carried out often do not get the same attention. These can be the 'Achilles heel' of a project or an operation. All personnel in the organization should have training and skills in change management. Four basic questions they ask are:

> *What* could go wrong?
> *How* could it affect others or me?
> *How* likely is it to happen?
> *What* can I do about it?

ELEMENT 8: INCIDENT INVESTIGATION, ANALYSIS AND ACTIONS

Effective incident reporting, investigation and follow-up are necessary to achieve improvement in safety and risk management performance. This provides the opportunity to learn from the incidents and to use the information to take corrective action and prevent recurrence. The following is a typical sequence:

> incident occurs;
> respond to the emergency promptly and positively;
> collect pertinent information about the incident;
> analyze all causes;
> develop recommendations and take appropriate remedial actions; and
> follow through on the effectiveness of the actions.

Most incidents (minor to medium) are investigated by the supervisor or team leader, who knows the people and the area. Major incidents normally require special investigation teams with the correct mix of specific expertise. Near miss reporting must be strongly encouraged by management. In this case we have an incident with no damages but the basic causes can still and should be determined. By providing solutions to these basic causes, future incidents with damages can be avoided.

The importance of incident reporting and investigation is clearly stated in the following paragraph:

> What is not reported cannot be investigated. What is not investigated cannot be changed. What is not changed cannot be improved and, therefore, will happen again.

Near miss reporting is particularly important from an awareness point of view. People tend to forget their training unless an incident happens to refresh them. A proactive industrial safety and risk management program will benefit from the near miss communication which is one of the very important components that makes a program successful.

ELEMENT 9: OPERATION AND FACILITIES INFORMATION AND DOCUMENTATION

Updated information on the operation and facilities, properties of materials handled, potential safety and risk management hazards, and regulatory requirements is essential to assess and manage risk to an acceptable level and to also ensure a safe and reliable operation. Management must ensure that systems are in place and being used by the whole organization that enable this element to be handled effectively. The following are some key topics that must be included:

> Management must ensure that the operation and facilities information and documentation system is managed effectively. Updating responsibilities must be clearly defined and understood by all personnel.

> Drawings and other pertinent documentation, such as records covering operation, maintenance, inspections and facility changes, must be identified, be accessible and kept updated at all times.

> Materials involved in operations must have their properties and potential hazards identified. It is essential that these are documented and communicated.

> Applicable regulations, permits, codes, workplace standards and practices must be identified, documented and communicated to those who require them.

ELEMENT 10: COMMUNITY AWARENESS AND EMERGENCY PREPAREDNESS

Community awareness is a key factor in maintaining public confidence in reliability and integrity of any operation or project. Emergency planning and preparedness is essential to ensure that in the event of an incident all necessary actions are taken for the protection of the public, company personnel, contractors, environment and assets. This element should include the following:

> *A system* with clearly defined responsibilities to ensure recognition and response to the

community's expectations and concerns about the operation or project. Ensure the system allows for open communication with the community.

> *Emergency response plans* must be documented, accessible and clearly communicated. These should not only be designed by the appropriate company personnel but must have major input from and review by the community emergency services, i.e., fire, police, medical, etc. Where there are other industrial complexes close by it is important to become involved in a mutual aid agreement.

> *Equipment, facilities and trained personnel* needed for emergency must be identified and readily available.

> *Simulations and drills* must be scheduled at appropriate intervals to provide a state of readiness and ensure continuous improvement for first-class response.

ELEMENT 11: PROGRAM EVALUATION AND CONTINUOUS IMPROVEMENT

The program effectiveness must be frequently monitored and evaluated to ensure that it is meeting the needs of the organization. The evaluations should provide key information and direction so that the program can be continuously improved. The following activities can ensure that this element provides the desired results:

> Assess the operations or project at appropriate frequencies (depending on complexity and risk factors) to ensure all of the

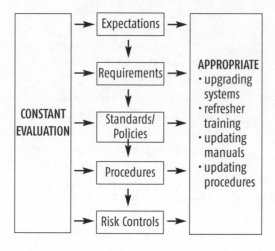

FIGURE 5–2: MANAGEMENT SYSTEM FOR
IMPLEMENTATION OF A SAFETY AND
RISK MANAGEMENT PROGRAM

elements of the safety and risk management program are meeting their objectives.

› Assessments should be conducted by multi-disciplinary teams with all the necessary expertise included.

› Recommendations from these assessments should be documented, evaluated and implemented as justified in order to ensure acceptable risk levels.

› The assessment process should be reviewed periodically to ensure the continuous improvement of this process.

Experience has shown that this element is quite often missed or not thoroughly implemented. It is absolutely essential to provide it with a very high priority in order to ensure that the whole program does not decay over time. *It is not sufficient to have the program designed and on paper; it must be practiced throughout the whole organization in order to obtain a safe and reliable operation.*

The first *duty of business* is to survive and the guiding principle of business economics is not the maximization of profit—it is *the avoidance of loss.*

—Peter Drucker

■ SUCCESSFUL PROGRAM IMPLEMENTATION

› *Installation* of a program in Safety & Loss/Risk Management has many pitfalls and requires a *lot of dedication* by whole organizations.

› Determine what is in place and is working as a first item—*use strength tree analysis.*

› Decide how large a program or how large an addition—use a team approach with representatives from the different areas of expertise.

› Design an installation time chart for good implementation—remember, organization is the key to carrying out *all of its regular duties.*

› More gradual installation is much better than the "big bang" installation.

Key Things to Do

› Give lots of thought up front.
› Determine what is in place.
› Decide what should remain.
› Decide on what to add—elements or complete program.

> Appropriate training for all personnel.

> Manuals (policy, expectations of all personnel, procedures, etc.).

> Roles and responsibilities.

> Stewardship.

> Continuous critique and stewardship.

Some of the Pitfalls

Start-up Effort Can Be a Problem

> Too much too soon.

> Not firmly imbedded into management objectives, systems and commitment.

> Some management expect results too soon—give up and then have a much more difficult time in restart-up. Usually this directive comes down from the top.

> Sound basic training effort quite large.

> Can ignore existing, effective partial programs already in place, causing negative acceptance.

> Best to have a reasonable sound *basic "safety"* program in place first.

Paperwork in Audits and Between Audits Can Be Large

> Probably, except for lack of management commitment, paper bureaucracy can be a major deterrent for effective implementation.

> Some very smart managers can use paperwork to cloud poor performance.

> Computerize and administer well—always verify data in the field by consulting with workers.

The Program and Auditing System Can Be Put in Place with Very Strong Support and Publicity

> Then runs down in energy until a few weeks before each yearly audit.

> Results never become significant.

> Top management can report to the CEO and board of directors that they are using the system— can get away with this type of performance for some time.

> When a major incident occurs, the poorly installed program can suffer some perceived credibility.

■ COST EFFECTIVENESS

In general all well *implemented* programs, whether they be small or large, are certainly cost effective. Not many companies analyze, or are able to analyze, the cost effectiveness of their programs. The results of the programs tend to be 'hidden' in such areas as lower maintenance costs, more efficient operations, longer run lengths, etc.

Usually companies do a good job of recording injuries, health incidents and environmental upsets, etc., and can tell if there is improvement from year to year. However, it is the hidden cost of not having a program that really can make a major difference in the company's bottom line. It is our experience that the less effective companies wait until they experience a major incident with high costs involved, or government fines/warnings before they really activate their program in a sincere manner. This reactive style of management is very costly and it is always easy to look back on an incident and determine how simple it would have been to avoid it had a good program been in place.

FIGURE 5-3: FIRST-CLASS COMPANY RESULTS

FIGURE 5-3: FIRST-CLASS COMPANY RESULTS

Typical Large Company First-Class Results with Integrated Safety and Risk Control Program

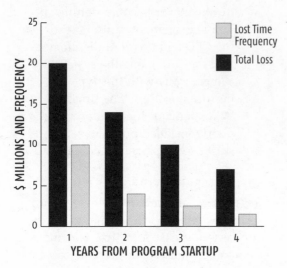

YEARS FROM PROGRAM STARTUP

Figure 5-3 illustrates the results of a study done on six refinery/chemical plants to analyze the effectiveness of the programs. Each of these plants had approximately 500 permanent staff and were located all over North America. It must be strongly emphasized that the programs were installed in a very sincere manner and monitored on a continuous basis by management throughout the years shown.

Refinery/Chemical Plants— Improved Results

A study was performed on six refinery/chemical plants to analyze the effect of an integrated risk management program. Figure 5-3 averages these results and clearly shows the benefits of both cash flow and people safety.

Please note the following definition of frequency rate.

FREQUENCY RATE: A MEASURE OF INJURY FREQUENCY

$$\text{Frequency Rate} = \frac{\text{Number of Injuries} \times 200{,}000}{\text{Total Exposure Hours}}$$

= Number of injuries per 200,000 hours worked

= Number of injuries per 100 person years

Example of Major Construction Project

It has been our experience that if an effective safety and risk management program is applied to a major construction project, the savings in costs, time, reliability and improved efficiency will certainly result.

The following project is a typical example:

COMPANY X	EXPANSION
• Total capital	$700 M
• Estimated "efficiency" savings	$ 80 M
• Estimated *minimum* savings due to excellence in safety and risk management	$ 10 M
• At 95% completion of the project, 4M man-hours were worked.	
• Injury frequency rate: 0.4	
• (Alberta Construction Average: 11.9)	
• Instead of 58,000* accidents in Alberta in 1986, this 0.4 rate would have reduced the accidents to approximately 5,000**	
• Major improvement in project manager's career.	

* all industry

** based on 0.4 versus *total* Alberta industry average of 5.6

CHAPTER 6

Safety and Risk Management for Project Managers

■ INTRODUCTION

Project managers have a tremendous responsibility for their project. They normally have a small personal staff, however, the main body of people report to them through a dotted line and also report to their own particular managers. For example, construction, design, operations and startup engineers all report back to their own management. This makes it very difficult for the project manager to always have strong influence over the personnel working on the project. It is important that the project manager pulls the whole team together through a first-class Safety and Risk Management program.

In our opinion, the best program to use for Safety and Risk Management is to take the integrated approach or PEAP (see Chapter 5).

■ THE INTEGRATED APPROACH TO RISK AND LOSS REDUCTION

"Best of Industry" Practices

Safety, Loss Management and Risk Management are all part of effective, efficient and continuously improving operations for project management and operations. ISRM must be part of all activities on a continuous, integrated and on-going basis. There must be a framework for managing to produce the right climate in the industry. Examples include the programs that Dow, DuPont, Shell, Imperial Oil, Syncrude, Bechtel, Kellogg, PCL and Ellis Don have developed. Imperial Oil has identified 11 elements of an overall program as part of its operations and project integrity. Imperial Oil believes these elements are essential for the *safe and reliable* management for all its operations.

Elements:

1. Management leadership, commitment and accountability.
2. Risk assessment and management.
3. Facilities design and construction.
4. Process and facilities information and documentation.
5. Personnel and training.
6. Operations and maintenance.
7. Management of change.
8. Third party services.
9. Incident investigation and analysis.
10. Community awareness and emergency preparedness.
11. Operations integrity assessment and improvement.

Each element is defined; has specific standards and objectives; and forms the basis for stewardship of performance.

■ THE IMPORTANCE OF SAFETY AND RISK MANAGEMENT FOR PROJECT MANAGERS

People Side of the Equation— Company, Contractors and Public (CCP)

The people side of the equation includes elimination of injuries and deaths, short-term suffering to victims and families. It also includes the protection of experienced well trained people and if handled well attracts the best employees. Items such as reduced absenteeism, turnover rate and improved morale are all outcome of good project management.

Assets (CCP)

A well designed and managed ISRM program will prevent damage to construction materials, machinery, mobile equipment, etc. It will also have positive effects on such items as spare parts inventory, project plans and techniques including special software. Public image and community relations will be enhanced.

Environment (CCP)

Environmental concerns includes protection of the air, water and land which is a must for any project manager. If damage is done to the environment major costs and law will ensue as well as the company's public image will be hurt.

Project Activity to Completion (CCP)

The project manager must pay particular attention to schedule, budget and productivity otherwise the manager will have problems with present and future customers. The long-term viability of the project could also be affected.

Poor project management can certainly cause negative results. On average in Alberta one-hundred people lose their lives in work-related accidents. Another fifty thousand serious accidents are reported each year. The total costs (direct and indirect) to industry are approximately two-billion dollars a year.

■ THE ROLE OF PROJECT MANAGERS

Project managers are project team leaders, inter-team facilitators, skilled people handlers and first-class leaders. They often can be equated to an orchestra conductor who has to combine the talents of many disciplines to achieve wonderful results.

The team members of the project managers' team can come from business, engineering, construction, services, operations and senior management. They can also include government and local authorities.

The project manager must continually push safety and risk management at all stages of the project in order to be on schedule, budget and meet quality standards. There is a saying in the industry that excellence in safety performance and risk management gives you "More Construction For The Money."

■ PROJECT MANAGERS' AREAS OF RESPONSIBILITY

Before Approval for Expenditures (AFE)
Project managers who are new to their role quite often do not realize that they have major responsibilities before expenditures are approved by executives on any particular project. This period is when project objectives, design basics, broad cost estimates, return on investments, and risk assessments are done. The project manager must be involved with all of these steps so that the presentation for project approval is well handled.

Engineering Design
Project managers must review basic designs well before construction is even thought of. They also should have responsibility in the area of review by senior management. Construction should not start to any degree unless the project manager has reviewed the detailed designs and has ensured risk studies have been carried out.

Construction
This is one of the major responsibilities of the project manager and must be carried out with strong emphases on safety and risk management. It includes contract selection, trades, man power quality assurance, materials management loss control and cost control. At construction completion there must be a testing of facilities as per design intent. The project manager must insure that the client accepts the facilities as built.

■ STEPS TO ACHIEVE EXCELLENCE BY COMPANY X

Safety Philosophy
> All injuries can be prevented.
> Safety is everybody's responsibility.
> Front-line supervisor has a particularly important role.
> Good performance will be rewarded.
> Failure to comply with safety requirements will result in disciplinary action.
> Senior management is totally committed and visible.
> All employees are responsible for their own personal safety.

Strategy
> One overall safety program for the project.
> Pre-screen sub-contractors' past performance.
> Pre-job meeting for specific contracts to review plans and potential hazards.
> Personal orientation for all employees.
> Continuous management audit.

- All incidents/near misses to be investigated and a report prepared.
- Give positive and negative feedback to sub-contractor and individuals.
- Safety program *must* involve everyone at site particularly watch for "short-term" people.

Results
- Total capital—$800 million over three-year project.
- 4-million-man hours—injury frequency rate 0.4.
- Efficiency savings due to Risk Management—$80 million.

■ SUMMARY
Project managers must ensure they have a first-class safety and risk management program in place. They also must ensure it is built into their total activities for the project. The program in chapter 5 can be used as a basis but it can also be modified to suit the size and complexity if the project. It is our experience that it is important to have senior management, middle management and contractors all agreeing on the content and the practice of an ISRM program.

CHAPTER 7

Risk Assessment, Analysis and Management

■ THE RISK MANAGEMENT PROCESS

Risk management is that part of the management process that deals with identifying, assessing and controlling hazards to people, the environment, assets and production (PEAP), and managing the risk that remains after risk controls have been put in place. The steps of the risk management process can be seen in Figure 7–1.

Referring to the symbols in Figure 7–1, the **System Description** identifies the components of the facility under study, how they operate, the inventory of hazardous substances and the surrounding area that might be impacted by events in the facility.

Review Requirements refers to management systems where specific reviews of facilities are carried out on a regular basis. These could be research reviews, project reviews, compliance reviews, insurance reviews and management directed reviews. All these reviews have a specific function to evaluate for hazards.

Hazard Identification addresses the question: "What can go wrong?" Potentially hazardous events are identified and characterized in this step. For example, in a flammable liquid storage facility, the realization that a release might lead to a pool fire and/or an explosion, and that there could be small, medium, large or catastrophic releases, constitutes the hazard identification step. Hazard identification may receive input from the ongoing monitoring programs through the recognition of factors or conditions that could initiate a loss. This step is often experience driven but must also adapt to new inputs. Hazard identification is really

FIGURE 7–1: THE RISK MANAGEMENT PROCESS

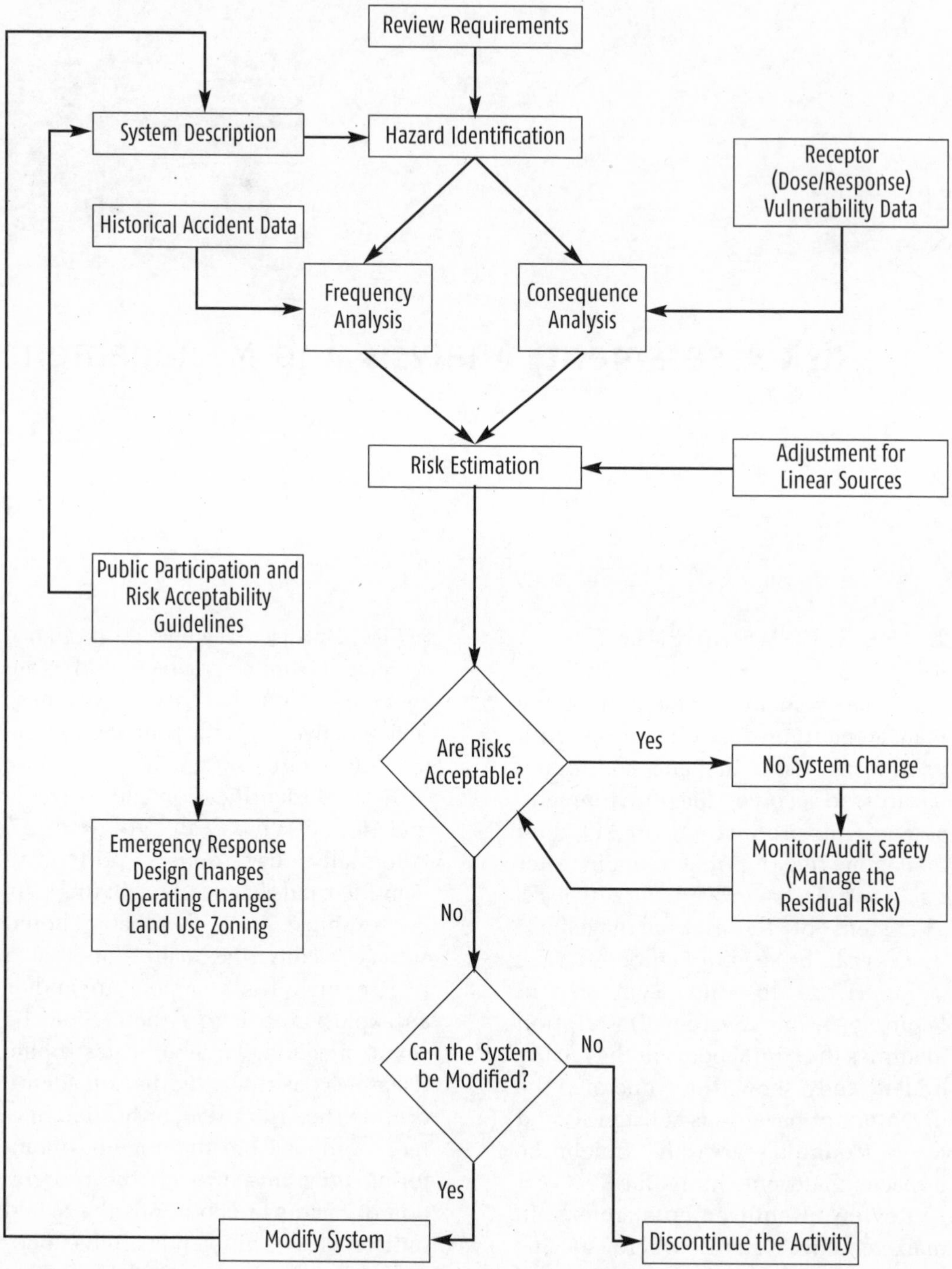

the trigger for the risk management process to begin.

Before risk can be managed, it must be understood. **Risk Analysis** defines the risk of a hazardous facility and the reductions in risk achievable given certain risk control measures. It addresses the questions: "How often is the event expected to occur?" and, "If it occurs, what are the consequences of the event?" Risk is a function of the likelihood of an unwanted event and the potential severity of its consequences, and may be expressed as:

Risk = a function of {frequency, probability, consequences}.

Frequency Analysis makes use of historical accident data, preferably from events in similar facilities. For rare events, fault and event tree analyses can be used to determine the scenario frequency based on equipment and human failure rates.

Consequence Analysis uses modeling methods to estimate the behaviour of the releases of hazardous substances and their impact on receptors making use of vulnerability (dose/response) data and models.

Risk Estimation is the process by which the frequencies and probable consequences of events are combined to quantify risk. The results of risk estimation are used extensively in risk management decisions throughout the world. The uncertainties in estimating the chance of rare events, and in projecting the effects on human populations, are considerable; however, high uncertainty does not mean high risk. Estimation of uncertainties in risk estimates is currently an area of active research.

Risk Evaluation is the stage at which values and judgments enter the decision process, explicitly or implicitly, by including consideration of the importance of the estimated risks and the associated social, environmental and economic consequences, in order to identify a range of alternatives for managing the risks. This is the process that answers the question: "Is the risk judged to be acceptable?" and, if not, "What do we need to do about it?" Whether an entity judges a risk to be small or large, acceptable or unacceptable depends upon many factors. Voluntary risks are those we assume due to some perceived benefit (e.g., smoking, skydiving, etc.). Involuntary risks are imposed on people by decisions made by others or by natural occurrence (e.g., secondhand smoke, violent storms, etc.). A hazardous facility is often seen as posing an involuntary risk on someone living nearby (especially if the facility is new). But it might also be seen as a voluntary risk if someone chooses to live near an existing facility, especially if the person is aware of the risks before moving there.

Risk Assessment is the process of risk analysis and risk evaluation. It is, in other words, the quantification and ranking of risks. It must precede the decision to mitigate or control a risk, otherwise, all risks would be treated as equal. Risk assessment is objective and user oriented. If the risks are judged to be acceptable, then further risk control measures or system changes will not be required; however, even for acceptable risk it is essential to develop programs to monitor the situation so that it does not deteriorate over a period of time.

Monitoring of the project/operation must be carried on continuously by all personnel at all levels in the organization to detect deviations and potential hazards. This can be done visually or by hi-tech sensing systems. Safety audits and inspections are two tools used for this purpose. If it is judged that further safety improvements are required, risk control options, including system modifications, can be examined.

Risk Control answers the question: "What can be done to reduce risks?" and accomplishes this by decreasing the likelihood and/or consequences of a hazardous event. Risk control measures will have certain costs associated with them. By estimating the risk reduction expected for each option, it is possible to assess the costs and benefits, and make informed decisions on which option should be selected. Risk reduction solutions could include:

> substitution in the process;
> change in the design of process systems;
> modification of control systems;
> organizational change;
> operating and maintenance procedures;
> personal protective equipment;
> improved communications;
> increased or varied training; and
> simulators to improve understanding.

The risk controls must be subject to monitoring to assure the risk remains acceptable. Risk controls mitigate the risk, but can never totally eliminate the risk.

Communication of the risk to the public exposed to these risks, **Public Participation** regarding the acceptability of these risks, and the **Risk Control** **Measures** that would be implemented are essential components of the **Risk Management Process**.

Application of the Risk Management Process

For a typical project, risk management must be carried out for all phases of the project's life cycle before major and significant changes are implemented. These include:

> initial project selection;
> process design;
> basic and detailed engineering;
> construction;
> commissioning and start-up;
> normal operations;
> plant additions and continuous improvement initiatives;
> plant shutdowns, maintenance; and
> plant demolition and site clearance.

Risk assessment is most appropriately carried out by teams with a variety of expertise. The disciplines represented on the team must reflect the requirements of the project/process being assessed. The team leader is extremely important to the success of the assessment. He or she should be trained in the methodology being used and act primarily as a facilitator to keep the team focused, provide discipline, emphasize objectivity and encourage participation from all team members. A heavy-handed leader can easily prevent the team from reaching the optimum solutions. It is very appropriate that decisions on final solutions be made as a team effort, e.g., manager, designer and user; however, it is also important to obtain "buy-in" at the working level, i.e., from the employees directly involved.

Action items that result from the above risk solutions should be prioritized and categorized. They must be assigned a completion date and have a person responsible. For major studies there can be many actions required. These actions can have different priorities for completion with short-, mid- and long-term end dates. In this case some form of project planning system may be required including an assigned project manager. Progress on the action items must be reviewed/stewarded by *senior management*.

Risk assessment, analysis and management should be on-going activities, as part of the overall safety and risk management program. There are many examples over the years where risk assessment and analysis could most likely have prevented a major disaster occurring, such as:

Flixborough (chemical plant)
Chernobyl (nuclear plant)
Lodgepole Blowout (gas well)
Exxon Valdez (oil tanker)
Bhopal (chemical plant)
Piper Alpha (off-shore oil rig)
Challenger (space shuttle disaster)
Hyatt Regency Hotel collapse
Syncrude Coker 8-2 Fire, 1984

Case studies on many of these incidences are included in Chapters 10 and 11.

There is no question that the process of risk assessment, analysis and management is easily justified since this is the core of proactive management; however, one of the keys is to match the scope of the studies to the potential risk.

■ HAZARD IDENTIFICATION AND RISK ASSESSMENT TECHNIQUES

There is a variety of risk assessment/analysis techniques that are currently used by industry. It should be understood that the following descriptions are not meant to be a full explanation of each technique or a complete listing of all of them. It is important that additional reading be done on the subject. Four examples of reference material are:

> A Manual of Hazard & Operability Studies, R. Ellis Knowlton. Chemetics International Company Ltd., Vancouver, B.C.
> "HAZOP and HAZAN" by Trevor Kletz, ISBN 1–56032–2764 (third edition).
> "Dow's Fire and Explosion Index, Hazard Classification Guide" (published by the American Institute of Chemical Engineers).
> "Dow's" Chemical Exposure Index (also published by the AICHE).

Different countries and industries tend to use dissimilar terminology or similar terminology with different meanings for this topic. Therefore care must be taken to assure that the participants in any study use and fully understand the basis for the method and the terminology that is used.

The following sections provide details of some of the techniques that are available for hazard identification and risk assessment. Figure 7–2 provides some guidance regarding the complexity (casual to structured) of the methodology and the time required to complete the study. This figure provides some guidance although the relative position on the complexity scale may change depending on the scope of the study.

Field Risk Observation

This type of assessment requires that employees constantly identify any change that may result in higher risk.

FIGURE 7–2: TYPES OF RISK ASSESSMENT/RISK ANALYSIS METHODS VS. TIME INVESTED

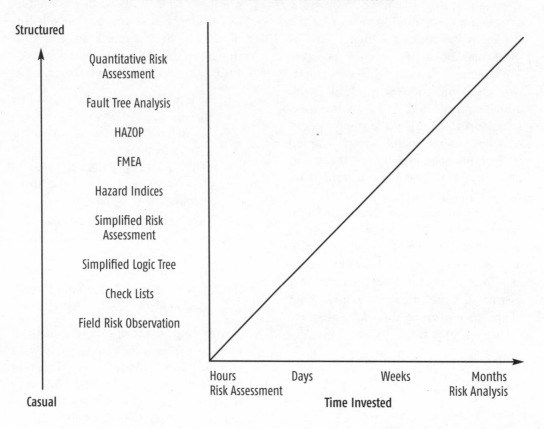

All employees should be given training in these observation skills. When doing their normal task or making any changes, they should ask the following questions:

> Why am I doing it at all?
> What could go wrong?
> Could it affect me and/or others?
> How likely is it to happen?
> What can I do about it?

Changes can take the form of an abnormal sound in the equipment, unusual odour around a facility, changes to procedures, practices or facilities. It is important to have safety "sessions" around this topic on a fairly frequent basis to keep people aware that risk assessment is a key part of their job. A system must be made available to ensure that these risk observations are reported and acted upon in a prompt manner.

Checklists

Checklists provide a means for ensuring that a plant, process or facility conforms to existing codes, standards of good practice, or company procedures. Employees using checklists are looking

FIGURE 7-3: SIMPLIFIED LOGIC TREE ANALYSIS FOR AN EXPLOSION

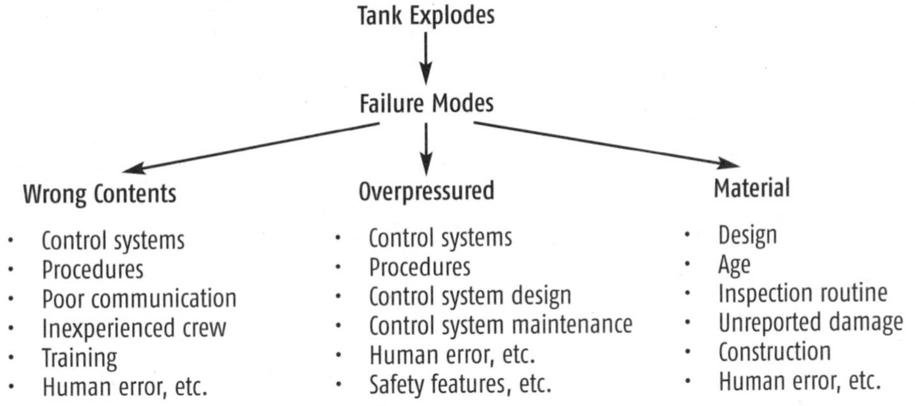

Tank Explodes

↓

Failure Modes

Wrong Contents
- Control systems
- Procedures
- Poor communication
- Inexperienced crew
- Training
- Human error, etc.

Overpressured
- Control systems
- Procedures
- Control system design
- Control system maintenance
- Human error, etc.
- Safety features, etc.

Material
- Design
- Age
- Inspection routine
- Unreported damage
- Construction
- Human error, etc.

for deviations that can increase risk. An example of this is a commercial airline pilot and crew checking the critical functions on the aircraft before take off. If there are any major deviations found, the aircraft is grounded until the problem is solved.

Checklists are only the beginning of risk assessment and are only as good as their quality and number of items included. Furthermore, a checklist can never be entirely complete or meet the needs of every situation. Care has to be taken to ensure that other pertinent items not included on the list are not overlooked, and management has to ensure that further hazard identification and risk analysis is carried out.

Simplified Logic Tree (what-if)

Logic trees provide graphic representations of the interrelationships of basic causes that can result in an undesired incident. Contrary to the other methods described here, it starts with effects (i.e., a potential situation) and works back to the causes in a top-down approach. The undesired incident, e.g., an explosion, is the starting point and the pathways of events leading to the incident are identified, see Figure 7-3.

Each branch and sub-branch is developed down to basic causes, the most probable cause pathway is identified and potential safeguards are evaluated. This analysis must be carried out by an experienced team that includes a cross-section of the organization. The team should consist of individuals with knowledge of all dimensions of the risk, e.g., managers, engineers, geologists, geophysicists, business people, operators and mechanics, and should include people with hands-on experience.

Semi-Quantitative Risk Assessment

Semi-quantitative risk assessment is based on the risk equation:

RISK = a function of
{Likelihood x Consequences}.

It is a simplified method that is almost always conducted by a multi-disciplinary team. This team should have a combined complete understanding of the system to be reviewed and must reach agreement on the following:

> a statement of objectives;
> a clear definition of the system(s) to be analyzed;
> design and operational details of the system;
> a listing of principal concern categories;
> a listing of all known assumptions and constraints;
> the time constraints that govern the risk assessment; and
> the personnel required (and available to support the risk assessment at various stages.

The team must have management's support and commitment to implement key recommendations that evolve from the study. Once a risk is identified, management has an ethical responsibility, as you do as a professional engineer, geologist or geophysicist, to reduce the risk to an acceptable level.

The assessment is done in a table form. Table 7–1 presents a risk assessment work sheet for a propane-filling depot under normal operation. Since this is completed before any particular incident, it addresses the scenarios that could potentially happen, their impact and their likelihood. It also includes the level of risk involved, the recommended controls and the residual risk after their implementation.

Under the headings "impact," "likelihood," "risk" and "residual risk" it should be noted that H, M, and L stand for High, Medium and Low. The team must assign these values based on their technical expertise and experience. In the "risk" column the result comes from multiplying the likelihood and impact assessment, e.g.,

$$M \times H = H.$$

Mostly a High (H) on the left hand side of the equation produces a High (H) on the right hand side. This is not always the case; the team must use their judgment. For example, a Low probability multiplied with a High impact can result in a Medium risk (L x H = M). In some cases, it may be necessary to quantify the values.

The controls or recommendations that need to be implemented can be determined by the team and/or additional experts based on the level of risk found in the study. These controls and recommendations must, when put in place, reduce the residual risk to an acceptable level. Residual risk is that level of risk that remains after all controls have been put in place. The determination as to whether this risk is acceptable should be determined by taking into consideration laws and regulations, society's values, and the company's objectives and standards. Table 7–2 provides a typical generic risk criteria as an illustration. The criteria shown are flexible and can be changed, if necessary, to suit the system that is being assessed.

This risk assessment technique is used extensively by industry. Although it is fairly simple in design it has proven very effective in evaluating the majority of risks in the industrial setting. It is an excellent tool and should be considered before using.

TABLE 7-1: SEMI-QUANTITATIVE RISK ASSESSMENT

SYSTEM: PROPANE FILLING DEPOT   SCOPE: NORMAL OPERATION

Item	Concern	Impact Rationale	I	Probability Rationale	P	Risk	Controls	Residual Risk
Propane Tank	Leak	Loss of inventory. Small fire.	L	Highly unlikely since tank is code built and tested.	L	M	Leak test system before commissioning.	L
	Catastrophic Failure	Explosion and fire causing injuries and property destruction.	H	Tank is protected from over-pressure by relief valve. Fire or external impact could damage tank.	L	M	Provide security barrier around tank. Post evacuation notices in event of fire.	L
Piping	Flow restriction	Inconvenience to user. May present fire hazard.	L	Debris or corrosion products in line. Possible ice plug.	M	M	Regular main-tenance. Provide heat tracing on line to prevent freeze up. Develop filling procedures.	L
	Leak	Loss on inventory. Small fire.	H	Piping subjected to abuse may develop leaks at connections.	M	H	Regular leak testing. Erect no smoking signs and remote isolation valve.	M
	Rupture	System depressured. Large flash fire possible involving the tank.	H	Highly unlikely if quality piping system installed. External object could strike piping.	L	M	Install bracing and shield around piping. Design regulator to quick shutoff if down-stream pressure drops rapidly.	L
Metering	Calibration error	Customer overcharged. Poor public relations.	M	Not likely given the frequency of refilling the storage tank.	L	L	Keep accurate records and calibrate system on a regular basis.	L
	Valve may freeze open	Cannot shut off system causing spill.	H	Unlikely if system is designed properly.	L	M	Trace circuit. Install emergency shutoff.	L
Customer	Spill propane on ground or on hot surface	Fire or explosion.	H	Possible but not likely.	L	M	Post operating instructions and hazard warnings. Install quick shut off.	L

I = Impact; P = Probability. L = Low; M = Medium; and H = High.

Source: Syncrude Canada Ltd.

TABLE 7-2: "GENERIC" RISK CRITERIA

RATINGS	IMPACT	PROBABILITY
At least once per year.	**High** P Disabling injury, loss of body part or fatality. E Reportable violation, toxic release. A High repair cost (Typically > $100 k). P*Loss of function of facility for an extended period, with business consequences, major quality deviation.	**High** • Repetitive event. • At least once per year. • Several times in the life cycle of a project. • Has happened frequently in similar circumstances. • Greater than 50% chance of occurring.
	Medium P Medical aid injury. E Non-reportable spill, line toxic release. A Moderate repair cost (typically > $10k). P*Short duration loss of function, serious quality deviation, medium business impact.	**Medium** • Infrequent event. • May only happen occasionally (less than once per year). • Has been observed in similar circumstances. • 10% to 50% chance of occurring.
10 to 50% chance of occurring.	**Low** P First aid injury. E Minor leak, nontoxic fugitive emission. A Low repair cost (typically < $10k). P*Brief interruption or minor quality deviation.	**Low** • Unlikely event. • Never happened to date. • Has never been observed but is still believed to be a possibility. • May happen less than once in 10 years. • Less than 10% chance of occurring.

Legend: P = People; E = Environment; A = Assets; and P* = Production — P* is shown to designate production rather than people

Note: Generic Risk Criteria should also consider other business losses, e.g., reputation damage, loss of confidence among stakeholders, e.g., bankers, insurers, regulators, public, employees and shareholders.

Hazard Indices (Dow's F&EI)

The Index method was originally developed to rank the relative loss potential for various plants and processing facilities. Dow Chemical's "Fire and Explosion Index" (F&EI) is one of the most popular index procedures, from which the Maximum Probable Property Damage (MPPD) can be estimated. The base MPPD is the dollar value of the equipment within the exposure area and the actual MPPD is the base modified by a credit factor for safety features designed into the process.

The Dow F&EI is especially useful where flammable, combustible or reactive materials are involved. It is not so well suited for facilities such as generating plants, office buildings, water treatment facilities and distribution systems. Other indices that are sometimes used are the Mond Index and the Dow Chemical Exposure Index. The process requires involvement from an experienced professional team to perform calculations such as:

> Process unit material factor (MF).
> General process hazards factor (Magnitude F_1).
> Special process hazards factor (Relative Probability F_2).
> Generally MF x F_1 x F_2 represents the Fire and Explosion Index.

Dow's Chemical Exposure Index (CEI)

Dow's Chemical Exposure Index uses a quantitative method. It is used to evaluate the risk associated with a potential release of toxic vapours into the atmosphere. This can be used to determine additional preventive measures. It is also useful for emergency planning, particularly if there is a danger to the public.

Failure Mode, Effects, and Criticality Analysis (FMECA)

A hazard can have several origins, and a detailed analysis of potential causes must be made. Reliability engineers often use a method called Failure Mode, Effects and Criticality Analysis (FMECA) to trace the effects of individual component failures on the overall failure of equipment. This type of analysis is equipment oriented instead of system oriented. In their own interest, many manufacturers perform a FMECA on products prior to introducing them to the market.

The FMECA may be important to the industrial safety and risk manager, particularly if the equipment is critical to the health and safety of the employees. A large benefit to the manager is to complete the study before an equipment failure takes place. The results may direct attention to critical components, and aid in the institution of an effective preventive maintenance program. Table 7–3 provides an example of a FMECA Record Sheet.

The Guideword Approach to Hazard and Operability Study (HAZOP)

The Guideword Approach is a systematic and thorough technique for identifying potential hazards and operability problems. If justified it can be applied to the analysis of a whole facility. The purpose of a HAZOP study is to identify all possible deviations from the design intent and understand the consequences these deviations can have on other parts of the process and/or facilities. This technique is normally applied to the detailed design in order to implement preventative measures. It is cheaper and easier to make changes on the design

TABLE 7-3: FMECA WORKSHEET

| Item | Component | Failure or Error Mode | Effect On | | Hazard Class | Likelihood of Failure | Detection Methods | Compensating Provisions and Remarks |
			Other Components	Whole Systems				

TABLE 7-4: THE GUIDEWORD APPROACH

Guideword	Application to Material	Application to Operations
No	No material is present	No part of the expected operation is achieved
More	A greater quantity of material than intended	A greater operation than intended
Less	A lesser quantity of material than intended	A lesser operation than intended
As well as	An additional component present	An additional operation beyond that intended
Part of	One or more intended components is missing	One or more desired operations is missing
Reverse	The logical opposite to the desired material	The logical opposite to the desired operation
Other than	A totally different material	A totally different operation

drawings than to make changes when the project is under construction or in operation. Accurate, up-to-date and complete piping and instrument drawings (P&IDs) of the process being studied are a necessity.

The study should be performed by a knowledgeable team (5–6 persons), consisting of people with different backgrounds such as engineers, geologists, geophysicists, maintenance and operations personnel. A structured brainstorming technique using a set of guidewords is applied to the design intentions to identify potential causes of deviations and the possible consequences that could result. The seven basic guidewords are given in Table 7–4 along with an explanation of their application. To obtain a quality product it is important that the team members understand the intention of the process being studied and agree on a terminology. The team will come up with a list of potential hazards that will form the basis for further decision making.

Adaptations to the guideword approach for HAZOPs have been developed to achieve varying objectives such as maximizing results from limited availability of team members and incorporating risk estimates through the use of a risk matrix such as that described previously.

The Guideword HAZOP approach may be applied at a systems level to first screen for high risk systems before being applied to components. In addition, experience documented through checklists may be used to replace or select relevant guidewords.

Fault Tree Analysis

One of the most widely used methods for conducting risk analysis is the fault tree. The fault tree is a graphic representation of the interrelationships of basic causes that can result in an undesired event. In a qualitative analysis, this network picture can assist in answering the question "How can this undesired event occur?" Should the

FIGURE 7-4: FAULT TREE DIAGRAM WITH *and* AND *or* LOGIC GATES

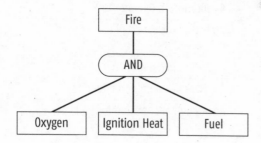

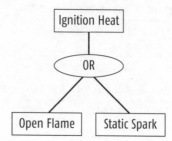

objectives also require a quantitative analysis, the fault tree is ideally suited to answering the questions:

> What are the chances of this undesired event occurring?
> Which causal factors are significant?
> How can the risks be reduced?
> What are the risk benefits from specified risk reduction measures?

Fault tree analysis is used to assess causes associated with an undesired event, to quantify the frequency or probability of this event occurring, and to analyze the risk sensitivity to changes in the frequency and/or probability of principal causal factors, for example, from implementing risk mitigation measures. Fault trees are particularly useful in the assessment of risks associated with industrial operations whose success or failure is dependent on the appropriate interaction of design criteria, process variables, equipment, control systems, management systems, operations and maintenance activities. They are a valuable

tool when trying to assess the risk of low-frequency/high-consequence events.

The fault tree provides a system model that is presented as a logic network of events that describe one or more of the above mentioned factors. Events constitute the basic building blocks of the fault tree and causal relationships between events are defined in terms of logic gates.

Two simple fault trees are shown in Figure 7-4 illustrating the use of the AND and the OR logic gates. The fault tree leading to a *Fire* includes the AND logic gate that may be described as follows: *Oxygen* AND *Ignition Heat* AND *Fuel* all have to be present at the same time for the top event *Fire* to occur. Similarly, the fault tree leading to *Ignition Heat* includes the OR logic gate that may be described as follows: either an *Open Flame* OR a *Static Spark* are sufficient for *Ignition Heat* to occur as described in the previous fault tree. Although there are many variations to these basic logic gates, the concepts of most fault trees may be described through the AND and OR gates as shown in Figure 7-4.

TABLE 7-5: COMPARISON OF ANALYSIS METHODS

Method	Methodology	System Complexity	System Understanding	Project Timing	Time Required
What if?	Bottom Up	Low	Medium to High	Anytime	1 hr or more
Checklist	Experience*	Low to High	High	Detail Design	1 hr or less
What if?/ Checklist	Bottom Up Experience	Low to Medium	Medium to High	Anytime	1 hr or less
FMECA	Bottom Up	Low to Medium	Medium	Detail Design	1 to 2 hrs
Event Tree	Bottom Up	Low to Medium	Medium	Preliminary Detail Design	1 to 2 hrs
Fault Tree	Top Down	Low to High	Medium to High	Preliminary Process Design Existing Systems	1 hr or less
Guideword HAZOP	Bottom Up	Low to High	Medium to High	Detail Design	3 hours
Structured Brainstorming	Bottom Up	Low to Medium	Medium to High	Preliminary Procedures	1 hr or less

* or existing documents, e.g., regulation, operation manual, etc.

When quantifying the fault tree, basic events located at the bottom of the tree are evaluated in terms of their probability or frequency of occurrence. The resulting value for the undesired system event, identified at the top of the fault tree, is calculated through the fault tree logic. As may be seen from evaluating the above simplified fault trees, the AND gate implies that two or more conditions have to exist at the same time in order for the resultant event to occur. Such logic shows that the likelihood of the resultant event from an AND gate is lower than the likelihood of the constituent events. In contrast, the logic associated with the OR gate shows that the probability or frequency of the resultant event increases with an increasing number of constituent events because the occur-

rence of any one or more of these constituent events will immediately cause the resultant event to occur.

Fault tree analysis is a very flexible technique that may be applied with varying levels of detail and accuracy depending on the risk analysis objectives. Care must be exercised to retain information on the uncertainty associated with each analysis. Although the technique is often applied to detailed, complex systems, it is also easily applied to simplify systems and this range in complexity was previously illustrated in Figure 7-2.

■ SUMMARY OF METHODOLOGIES

There is a range of methods available to identify hazards and assess risks. It is up to the leader to determine the most effective methodology based on a number of considerations. Table 7–5 provides some guidance in selecting from the range of methodologies.

Methods to Assessing Risk

This chapter provides powerful "methods" to assess risk in design, construction, operations, etc., in a very logical fashion. These "methods" can also be used to determine the causes of any particular incident that has already happened.

Some of the benefits of these methods are as follows:

1. Provide a very logical and sequential pathway to the solutions although in most cases they require judgmental input.
2. When used to justify a course of action to senior management, they provide a very valuable and convincing opinion and certainly helps in reaching a final decision.
3. The layout of the risk assessment can easily be used to communicate to other groups affected by the final decision but not involved in the actual analysis.
4. In the future when changes are made to systems, facilities, etc., that have had a previous risk assessment carried out on them, it makes it very easy to go back and use the previous risk assessment as a reference. Then an additional risk assessment can be done for the changes.

Some of the things to look out for when using these methods:

1. They are not exact mathematical models. Some engineers tend to believe they are. They should apply experience and insight in order to reach their final decision.
2. Figure 7–2 shows the type of risk assessment/analysis methods versus time invested to carry them out. The Y-axis goes from the simple field risk observation all the way up to the complex quantitative risk assessment. The people doing the assessments tend to jump right up to the more complex methods before using the simpler ones. This is usually because of their training as engineers, business managers, medical doctors, etc. They have the feeling that the more complex the method they use, the better would be the result. This most often is not the case. In our experience 80% of the risk assessments required can be fully done by using the Simplified Risk Assessment technique. This is halfway up the Y-axis. It saves a lot of time and money. Obviously there are places for the more structured or exotic techniques to be used. However, the golden rule is, where possible, the simpler techniques should be more often used and can certainly do the job.
3. Sometimes "new" methods are published in the various papers. One has to be careful that they perform as well as the ones we have explained here. Note that newer methods will come forward that could prove better.

CHAPTER 8

Causation Model and the Importance of Systematic Incident Investigation

" Prescription without diagnosis
is malpractice, whether it be in
medicine or management"

—Karl Albrecht

A thorough understanding of incident causation is critical to the development of appropriate preventative measures. Managers who believe that most incidents are caused by carelessness are likely to resort to disciplinary actions or incentive programs. When this direction is taken, most often the incident problems are covered up or concealed rather than solved. Professionals must have a deeper understanding of the underlying (basic) causes of incidents. A framework for analyzing the sources of the incidents and control their effects is provided in this chapter.

Individuals committed to safety and risk management believe that almost all incidents can be prevented; natural disasters are the major exception. Most people use the word "accident" for an undesired event. However, the use of the word "accident" takes away the responsibility or the event, i.e., "it is a matter of luck". The more effective term is "incident". The following is the accepted definition:

Incident: An undesired event that does or could result in injury to people, damage to the environment or loss of assets and/or production (PEAP). This includes both an actual loss or a near miss.

An incident leading to a loss is most often the result of contact with a substance or a source of energy (mechanical, electrical or thermal) above **the threshold limit** of the body or structure involved or the environment.

TABLE 8-1: EXAMPLES OF IMMEDIATE CAUSES

Substandard Practices	Substandard Conditions
• Operating equipment without authority	• Inadequate or improper protective equipment
• Failure to follow established procedures	• Defective tools, equipment or materials
• Making safety devices inoperable	• Fire and explosion hazards (hidden)
• Failing to use personal protective equipment	• Poor housekeeping, disorderly workplace
• Servicing equipment in operation	• Hazardous environmental conditions
• Under influence of alcohol, drugs	• Inadequate training, expertise, etc.

The occurrence of the incident is controllable. However, the severity of the incident (consequences or results) is less controllable and is often a matter of chance. It is imperative to be proactive in controlling both occurrence and severity with stronger emphasis on the former. Table 8–1 shows examples of injuries to the body where energy exceeded the threshold limit. Similar examples can be readily found for exceeding the threshold limits of buildings, structures, facilities, equipment and environment. This type of data is valuable in determining proactive solutions to prevent incidents happening.

■ IMMEDIATE CAUSES

Immediate causes are the circumstances that immediately precede and can also develop during the incident. They most often can be identified readily at the beginning of an incident investigation. A typical classification and examples of these immediate causes are given in Table 8–1. Table 8–2 shows some examples.

■ BASIC CAUSES

Basic causes are the underlying causes for an incident and their discovery required in-depth search. The basic causes are the reason why the substandard practices and conditions exist. These causes provide the factors that when identified permit meaningful management action to mitigate and avoid additional incidents.

In this process there are three questions to ask in order to identify the basic causes:

1. Why did that substandard *practice* occur?
2. Why did that substandard *condition* exist?
3. What *failure* in our supervisory/ management system permitted that practice or condition?

It is very important to keep asking the question "why" or "what" after each answer until the final basic causes are reached:

> Why did the worker perform a substandard practice? Because of lack of training.
> Why? Because the budget for training was cut.
> Why? Management did not see it as a priority.

TABLE 8–2: INJURY CAUSATION TABLE WITH EXAMPLES AND COMMENTS

INJURIES CAUSED BY DELIVERY OF ENERGY IN EXCESS OF LOCAL OR WHOLE-BODY INJURY THRESHOLDS		
Type of energy delivered	**Primary injury produced**	**Examples and comments**
Mechanical	Displacement, tearing, breaking and crushing, predominantly at tissue and organ levels of body organization.	Injuries resulting from the impact of moving objects such as bullets, hypodermic needles, knives and falling objects and from the impact of the moving body with relatively stationary structures, as in falls and plane and auto crashes. The specific result depends on the location and manner in which the resultant forces are exerted. The majority of injury is in this group.
Thermal	Inflammation, coagulation, charring and incineration at all levels of body organization.	First- second- and third-degree burns. The specific result depends on the location and manner in which the energy is dissipated.
Electrical	Interference with neuromuscular function and coagulation, charring and incineration at all levels of body organization.	Electrocution, burns, interference with neural function as in electroshock therapy. The specific result depends on the location and manner in which the energy is dissipated.
Ionizing radiation	Disruption of cellular and sub cellular components and function.	Reactor incidents, therapeutic and diagnostic irradiation, misuse of isotopes, effects of fallout. The specific result depends on the location and manner in which the energy is dissipated.
Chemical	Generally specific for each substance or group.	Includes injuries due to animal and plant toxins, chemical burns, as from KOH, Br_2, F_2, and H_2SO_4 and the less gross and highly varied injuries produced by most elements and compounds when given in sufficient dose.

Source: Bird Jr., F.E. and Germain, G.L. (1992). *Practical Loss Control Leadership*, Loss Control Management, Det Norske Veritas, Inc.

TABLE 8–3: EXAMPLES OF BASIC CAUSES

PERSONAL FACTORS	JOB FACTORS
• Inadequate physical/physiological capability	• Inadequate leadership/supervision
• Inadequate mental/psychological capability	• Inadequate engineering
• Physical or physiological stress	• Inadequate purchasing
• Mental or psychological stress	• Inadequate maintenance
• Lack of knowledge	• Inadequate tools and equipment
• Lack of skill	• Inadequate work standards

In response to these questions, management must restore the priority and thoroughly understand how their decisions affect the safety and risk management performance. A typical classification and examples of these basic causes are shown in Table 8–3.

■ LACK OF CONTROL (STEWARDSHIP)

Control is one of the four essential functions of management. That is, they have to plan, organize, lead and control (steward). All of these functions relate to any manager's function regardless of work or title. In all of these steps they should integrate safety and risk management principles and practices throughout their total activities. There are three common reasons for lack of "management" control. These are:

> inadequate program;
> inadequate program standards; and
> inadequate compliance with standards.

Problems and loss producing events are seldom the result of a single cause. Most often the incidents happen because of **multiple causes**.

Conclusions

> Incidents with consequences for people, environment, assets and production are *caused*. They do not just happen.
> The causes of loss can be determined and proactively controlled.
> There are very large incentives for carrying out safety and risk management activities on a continuous, proactive and aggressive basis.
> Causes for any particular incident are normally multi-factorial.

■ IMPORTANCE OF SYSTEMATIC INCIDENT INVESTIGATION

The importance of incident reporting and investigation cannot be overstated. The basic causes have to be determined in order to provide the key solutions that will prevent future similar incidents. The following lists indicate some of the key reasons for reporting and investigation incidents:

Why Report Incidents?

> What is not reported cannot be investigated and recommen-

FIGURE 8-1: INCIDENT CAUSATION MODEL

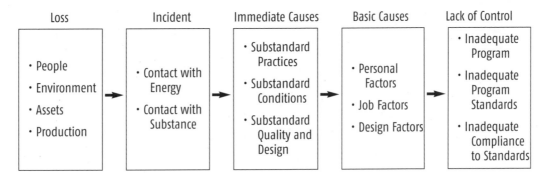

Loss	Incident	Immediate Causes	Basic Causes	Lack of Control
• People • Environment • Assets • Production	• Contact with Energy • Contact with Substance	• Substandard Practices • Substandard Conditions • Substandard Quality and Design	• Personal Factors • Job Factors • Design Factors	• Inadequate Program • Inadequate Program Standards • Inadequate Compliance to Standards

Note: The above model is based on model developed by Bird Jr., F.E. and Germain, G.L. (1992). *Practical Loss Control Leadership*. Loss Control Management. Det Norske Veritas, Inc.

dations cannot be made to prevent future events.

> According to provincial legislation in most provinces in Canada it is legally required that incidents involving bodily harm resulting in a worker being admitted to hospital for a specific period of time are reported to appropriate government department. In Alberta, if a worker is in hospital for more than 2 days, it must be reported to Alberta Labour.

> From the reporting, a database of loss history can be developed for future trend analysis.

> The reporting of minor incidents can identify problems in the operation before they lead to major incidents.

Why Investigate Incidents?

> To recognize the substandard practices and conditions that caused the incident.

> To identify the management system that failed to prevent it from happening.

> To recommend remedial actions that will prevent it from happening again.

The systematic incident investigation provides the basis for the basic cause analysis. The investigation contains the description of the top event (loss and incident) and makes it possible to determine contributing events (immediate causes) that could have caused the top event. Each of the contributing events are analyzed in order to determine its respective causes and the analysis is continued until the management system cause is determined (basic causes and lack of control). Figure 8-1 shows an incident causation model.

Making the appropriate recommendations is the most important part of an incident investigation report. The investigation is wasted if only facts are stated and conclusions are drawn on

FIGURE 8–2: INCIDENT INVESTIGATION FLOW PATH

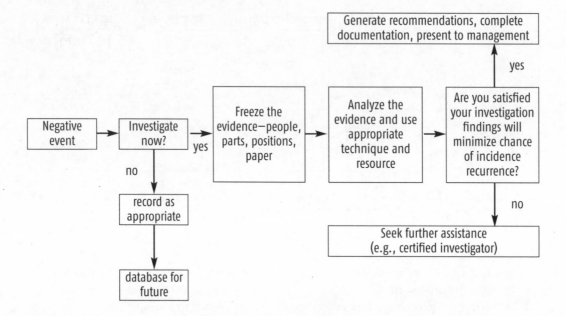

how the incident happened. Corrective actions are needed to avoid future incidents and specific recommendations should be made to address each of the basic causes that were established through the analysis activity. From the recommendations an action plan should be developed and responsibilities for the actions delegated to personnel who are held accountable for their completion. The process should be stewarded and recorded in action log that reflects agreed upon actions.

Not all the incidents must be investigated immediately. This is a judgment call made by the person responsible for conducting investigations. However, an incident should be investigated as soon after the incident as possible. If there are personal injuries involved, it has to be investigated according to the provincial legislation. However, even if the incident seems insignificant, it should still be recorded for future trend analysis. If it is chosen to investigate immediately, the evidence should be frozen (i.e., maintained) and analyzed. The investigation findings have to be evaluated to ensure they will prevent recurrence of future incidents and can provide the basis for recommendations. If that is not the case, assistance may be needed from certified investigators. Figure 8–2 diagrams an incident investigation flow chart.

■ FRAGILITY OF EVIDENCE

Right after an incident there can be major confusion. At this time important evidence can be damaged or deteriorated. In order to get the full benefit of and a high quality investigation, the site of the incident or event must be secured (i.e., frozen) within minutes.

FIGURE 8–3: PRESERVATION OF EVIDENCE

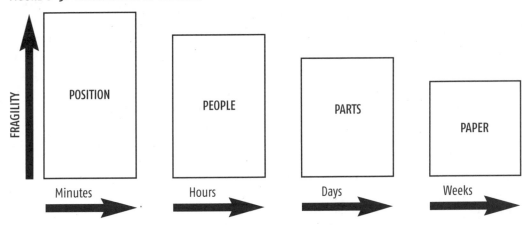

This will ensure that the position of people, material and debris is not removed. People who witnessed the event should be interviewed within hours and before there is any information sharing between the various witnesses. Larger parts involved in the incident will usually be preserved for several days after the event. Paper and computer logs documenting the process and work prior to and during the event will stay available for weeks. Figure 8–3 shows the fragility of evidence as a function of time.

■ INTERVIEWING TECHNIQUES AND HELPFUL HINTS

When interviewing an employee who has witnessed an incident, it is important the interview is conducted on a "one-on-one" basis. Physically, the interview should take place in a neutral and private setting and not in a superior's office since it may cause some reluctance to volunteer information. The interviewer has to make an effort to depersonalize the interview in order to

maximize the benefits. Open-ended and close-ended questions can be used to gather and confirm information. It is important that the interviewer is concentrating on listening and recording and does not make comments nor make attempts to direct or influence the interviewee. The interviewer should encourage the witness to return with any remaining follow-up if he/she remembers other details related to the incident. It should be emphasized that any detail could be of great importance. Interview as many eye witnesses as possible and compare their responses.

■ ROOT CAUSE ANALYSIS

Root cause analysis is the process for discovering and analyzing the very basic factors that allowed the negative event to happen. Often investigators find the immediate causes and do not go further. The basic causes are those that allowed the immediate causes to develop. They are more difficult to identify and are often not evident until after an incident has been thoroughly researched and investigated. One has to keep asking the question "why" after each individual cause has been identified. For example, if one of the causes of an incident was due to poor operator training, the investigator then has to ask the question "why". It could be due to a substandard training program. Again asking "why", this could be due to a failure of management supplying sufficient funding. This "why" questioning should go on until the investigator reaches the root or basic cause. This allows the investigator to solve for basic causes, which is much more important than solving for immediate causes. Most risk analysis systems have this process built into their logic but sometimes the person or team doing the study do not go sufficiently deep.

■ SUMMARY

When you are doing an incident investigation you will find the information presented in this chapter very helpful. The causation model works through questions that can identify cause. The systematic incident investigation information adds logic to your investigation and provides a stable and progressive basis for accurate results. It is fine basic information for the incident investigator.

CHAPTER 9

Human Factors

■ INTRODUCTION

The study of human factors and the resulting models and techniques that can be used to successfully improve the performance in industrial safety and risk management is now recognized as integral to risk management. Most of the leading companies in industry are now heading in this direction. They believe that by adding the study of human factors to their existing safety and risk management programs they can obtain significant improvements in the results they have already achieved. It is important to understand that the human factor method and techniques should be totally integrated with all elements of the overall safety and loss management program. It should not be a single and separate issue.

It has been acknowledged that "human error" occurring in design, operations and management is responsible for increasing the level of risk in all types of industries. Examples where human error was a major contributor are incidents such as Flixborough, Exxon Valdez, Piper Alpha and Phillips 66. (Some of these incidents are explored in case studies in Chapter 10.) Data from the Marsh and McLennan study "Large Property Damage Losses" (a thirty-year review, 1962–1991, which has not been superceded) indicates that human error was the main cause of 20% of the 170 major incidents occurring over the period. The average property damage loss in these human error cases was approximately $51.8 million each. In contrast only 4% of these major incidents were caused by design error, although their average loss value was $57.6 million each. Human factors are capable of being managed so that their contribution to risk and recurring losses is substantially reduced.

Management and the professionals (i.e., engineers, geologists and geophysicists) must take the leadership and provide the commitment to ensure that human factors play a significant role in their industrial safety and risk management program. Unfortunately, management often acquires the attitude that human error is a problem limited to the personnel at the "work face" level. They often believe that if the workers were better motivated then the problems with human factors would be greatly reduced. This causes management to use motivational campaigns to address the problem and, if this fails, disciplinary action will be taken against the worker instead. Management must understand that even motivated employees can perform "unsafe acts" and that the responsibility for creating the environment, conditions and the culture for minimizing human error rests squarely with management and the professionals. The current view is that most errors leading to minor or major losses are caused by factors that are mostly outside the control of the operator. Examples of such factors are the design of facilities, procedures, training and employee ability/work requirement matching. Management policies and practices can certainly optimize these factors. However, for these policies and practices to work effectively it is necessary that management change from their traditional blame and discipline philosophy to the human factors approach. That is, the management systems and culture can cause a large number of "human errors".

The willingness of management and professionals to change to this human factor philosophy is the first step. However, the implementation of the required processes and the integration of the processes into the safety and risk management program is the difficult application step. It requires significant knowledge and depth regarding causes of errors and appropriate remedial actions. The implications of human error for system reliability has been studied for a number of years. The "traditional" approach to human reliability was concerned with quantifying the probability of operator error when performing prescribed tasks. This is still necessary to provide input to the formal risk assessment. A wide range of techniques have now been developed to perform such analyses (see Chapter 7). However, this traditional method of investigating human reliability is limited in that it does not specifically address error reduction or the analysis of the underlying causes of errors.

The modern approach view error reduction as a primary objective and the role of quantification is to provide input to the cost-benefit analyses when choosing between alternative error reduction strategies. Most importantly the operator or worker is viewed as an **active problem solver** rather than a passive component in the process. This leads to a more systematic consideration of higher level cognitive activities such as problem solving, diagnosis and decision making that characterizes the operator's or worker's role in modern industry.

One of the key ingredients in this new human factor approach is the continuous development of an appropriate **organizational culture**. Management commitment to this new philosophy, including provision of systems and resources, is very necessary. However, there must be a willingness of the whole organization to participate in all the appropriate activities. They must provide the necessary input to the technical methods, such as task and error analysis. Active participation in and "ownership" by the workforce of the processes addressing the human factor issues is the major key to success.

Inside the scope of human factors, emphasis must be put on the fact that all employees do not usually perform at 100% of their efficiency at all times (they are only human). This must be taken into account when facilities are designed, constructed and operated. This should include the addition of such items as fail-safe devices, redundancies in controls, special procedures and practices, improved training methods, teamwork, etc. All of this is put in place to reduce the number of error provocative situations to an acceptable level. However, the workforce should still understand that they must still provide an acceptable and continually improving competency to the work process. An example of an error provocative situation, investigated in the United States is summarized:

> Six infants died in the maternity ward of the Binghampton, New York, General Hospital because they had been fed formulas prepared with salt instead of sugar. The error was traced to a practical nurse who

had inadvertently filled a sugar container with salt from one of two identical, shiny, 20-gallon containers standing side by side under a low shelf in dim light in the hospital's main kitchen. A small paper tag pasted to the lid of one of the containers bore the word "Sugar" in plain handwriting. The tag on the other lid was torn, but one could only make out the letters "S..lt" on the fragments that remained. As one hospital board member put it "Maybe that girl did mistake salt for sugar, but if so we set her up for it just as surely as if we'd set a trap..." (personal communication to the author)

When a system fails it does not fail for any one reason. It usually fails because *the kind of people* who are trying to operate the system, with *the amount of training* they have had, are not able to cope with *the way the system* is designed. This includes using possibly outdated procedures that they are supposed to follow *in the environment* in which the system has to operate.

■ WHAT ARE HUMAN FACTORS?

Human Factors addresses the interaction of people with other people, with facilities and with management systems in the work environment. **It identifies factors that affect human performance.** It provides practical ways to help reduce incidents while improving productivity (see Human Factors Model in Figure 9–1).

When people adjust their behaviours or environments to accomplish

TABLE 9–1: THE HUMAN FACTORS MODEL

Human Factors can influence results in many areas. These figures depict the "spectrum" or range of areas where Human Factors should be considered, with examples of actual human factors "in action".

SPECTRUM EXAMPLES	EXAMPLES
Workplace Design Facility layout Workstation configuration Accessibility	Easy accessibility to safety/critical and frequently used equipment is important in reducing the likelihood of human error.
Equipment design Controls (valves, handwheels, switches, keyboards) Hand tools Control systems	Easy accessibility to safety/critical and frequently used equipment is important in reducing the likelihood of human error.
Work Environment Noise Vibration Lighting Temperature Chemical exposure	Equipment used in extremely cold environments should be designed to accommodate increased clearance and decreased dexterity and force for workers who are wearing protective clothing.
Physical Activities Force Repetition Posture	Physically demanding activities should be designed to reduce the chances of injury.
Job Design Work schedules/Workload Job requirements vs. peoples' capabilities Behaviour-based safety Task design	Behaviour-based safety management observes work processes to identify and eliminate undesirable behaviours.
Information Transfer Labels/Signs Instructions/Procedures Communications Training Decision making	Hand-written labels and signs can indicate the need for clearer, more consistent information, which can reduce operator error.
Personal Factors Stress/Fatigue Age/Culture Boredom Motivation Fitness/Body size/Strength	Designs should accommodate personal factors, such as varying body size and strength.

FIGURE 9–1: HUMAN FACTORS MODEL

Operating Environment and Culture

FACILITIES
Pumps, control systems, panels, valves, cranes, etc.

MANAGEMENT SYSTEMS
Procedures, risk assessments, incident investigations, training, etc.

PEOPLE
Human characteristics and behaviour

their work, the changes are often inconsequential. But sometimes changes create new challenges. To save time or reduce discomfort, people may behave in risky ways. They may work in awkward positions, for example, or select available but improper tools. Attempts to follow unclear instructions, labels or warnings also increase chances of error and operational incidents.

Human Factors uses scientific knowledge from many disciplines— e.g., industrial engineering, psychology, physiology—to reduce incidents and improve operating effectiveness. It helps make tools and equipment more user-friendly by improving accessibility, equipment layout and presentation of information. It also examines communications systems and worker habits and recommends processes for encouraging safe behaviour.

Human Factors also considers human differences. The workforce is becoming increasingly diverse in physical characteristics such as strength and size. Differences in cultural background, education and experience are also increasing. Workers have varying capabilities and limitations in knowledge, problem-solving skills and speed. All these factors need to be considered in the design of facilities and management systems and are areas in which Human Factors technology can help.

Management Responsibility and Influence

In human factors it is important that management set the right atmosphere and continually provide a positive influence on it. Engineers in particular tend to deal with physical equipment, particularly in the early part of their career. They must understand the very high importance of human factors in any operation. Often we have seen equipment designed in a perfect engineering manner but with no human factors input to the point that this equipment cannot be operated by a human being, i.e., valve, pump, stairwell location, etc. Management must strongly encourage feedback from the working level on any changes, additions, new procedures, etc. It has always amazed us over our careers how much significant ideas and input can come from the worker that allows for a much more efficient and humane operation.

CASE STUDY—APPLICATION OF HUMAN FACTORS TO A PROJECT

Sable Offshore Energy Project

Issue

In 1996 the owners of the Sable Offshore Energy Project (including Imperial Oil Limited and Mobil Oil Canada Properties) agreed to proceed with a large-scale onshore and offshore development located in the Nova Scotia offshore region. As the pioneer gas development in the area, Sable provided a rare opportunity to incorporate human factors engineering (HFE) into the base design and philosophy of a new operation.

Actions

The endorsement of the senior management committee was obtained for the HFE program and budget. A separate HFE budget allowed the use of HFE professionals in the preliminary work without impacting construction budgets. The Sable HFE program was based upon seven key principles.

1. Involve HFE early in the project.
2. Assign an HFE champion.
3. Locate capability in engineering departments.
4. Base program on accepted HFE design standards.
5. Complete activities.
6. Design facilities such that human error is eliminated or minimized and seek to mitigate errors that may occur.
7. Extend influence of HFE beyond facility design

Implementation

› The role of HFE champion was assigned to the Senior Project Engineer.
› Work instructions outlining HFE expectations were issued to project staff.
› Technical staff training started immediately after program endorsement.
› HFE professionals were included as part of the engineering team rather than the SHE group.

Impact

The majority of HFE changes necessitated rearrangement of components. As they were incorporated early in the design process these were accomplished at minimal cost. Examples of component rearrangement based on HFE review are:

› Offshore heat exchangers were relocated to allow for improved crane access to facilitate bundle removal.
› Orientation and evaluation of large valve components located in a subcellar deck were changed to facilitate maintenance.
› Deluge piping was simplified and relocated because it interfered with component removal and blocked an access route.
› Access to valve stations at the gas plant dehydration system was simplified by eliminating multiple platforms and levels.
› Air cooler access platforms were modified to provide space for maintenance activities.

Human Factors was considered in procedures development, training, labeling and signage to enable efficient and effective training.

The HFE program introduced a number of standardized designs for the project. An example was developing a standard ladder design, specifically covered by one of the HFE guidelines.

Results

Sable HFE costs reflect personnel charges only. HFE driven design changes were considered design development. The original estimated cost for the HFE program was 0.07 percent of the facilities budget of 1.4 billion Canadian dollars. The actual HFE cost for this project was approximately half of the estimate.

■ CASE STUDY—INFORMATION DISPLAY IMPROVEMENTS IN A CONTROL CENTER

Imperial Oil Limited

Issue

Imperial Oil Limited (IOL) has a contract with a gas utility to deliver a specific volume of gas within a 24-hour period. The utility pays IOL a monetary incentive for accurate delivery. If the delivered volume is less than contract, IOL pays a penalty; volumes exceeding the contract are not reimbursed to IOL. Accurate delivery volumes are therefore critical to profitability.

The process control system in IOL's gas plant provided control room operators (CROs) with continuous date on supply rates from numerous gas sources. The CROs had to calculate this data to estimate total supply rates and cumulative volumes that would be delivered by the end of each 24-hour period. These estimates were subject to error, especially when supply rates varied.

Action

Improve information provided by the gas plant process control system to help operators accurately meet gas delivery contracts.

Implementation

IOL reprogrammed the process control system to calculate and display:

> supply rates from each source;
> total instantaneous supply rates from all sources;
> cumulative volume delivered from all sources; and
> supply rates required to meet contract demands.

Results

The information display changes increased the availability of useful information to the operator, reduced costs by $200,000 per year and improved relationships with IOL gas customers.

■ CASE STUDY—WORKPLACE ERGONOMICS PROGRAM IN OPERATION

Torrance Refinery Ergonomic History

Issue

In 1997, heritage Mobil's West Coast operations (which included refinery, terminals, pipeline and marketing groups) decided to develop a program to reduce the risk of ergonomic injury to employees.

Actions

> An Ergonomics Policy and Procedures manual was developed. The Ergonomics P&P includes procedures for design of new or modified workplaces and tools; reporting ergonomic injury symptoms; and for workplace assessments. It also provides training methods and defines roles and responsibilities.

> Ergonomics training was developed. Training is delivered through a custom video, a computer-based-training module, safety news-letters and supplemental classroom training for high-risk groups.

> A standardized method for workplace ergonomics assessments was developed.

> Pre-employment Agility Testing, Return to Work and Light Duty programs continued to be followed.

Implementation

> All employees received initial and annual refresher ergonomics training in the training format applicable to their group.

> Supplemental training was used for high-risk groups.

> Ergonomic Assessments were completed by Industrial Hygienists for 150 workplaces. In many cases, workplaces or work practices were modified as a result of ergonomic assessment. Assessments were completed for office environments as well as refinery process units, gasoline delivery tasks and *On-The-Run* food preparation tasks.

Results

> One example of the results of this program occurred in the Torrance Refinery in 1999. At the beginning of 1999, 31 employees who had some exposure to repetitive work had reported symptoms. At follow-up six months later, 20 of the 31 employees (65 percent) had become symptom-free. Six employees (19 percent) were improved but had minor symptoms. Two employees had not improved and one employee had worsened.

Conclusion

For the 31 cases of self-reported symptoms thought to be related to the work environment, 84 percent were symptom-free or improved with the ergonomic interventions undertaken.

The number and severity of ergonomic problems were reduced through a comprehensive program that included employee training and intervention at the first sign of symptoms related to repetitive work.

■ CASE STUDY—HUMAN FACTORS APPLICATION BY A PLANT MAINTENANCE GROUP

Exxon's Electrical Distribution Work Group

Issue

The Electrical Distribution Work Group at heritage Exxon's Baton Rouge chemical plant experienced a number of incidents that resulted in serious injuries and unplanned outages. The most serious incident occurred while two employees were attempting to test a breaker in a power distribution substation. A fault occurred, and an intense electrical arc caused a fire that destroyed the substation. The employees required medical treatment and were unable to perform normal duties during recovery. This incident and others led this work group to seek ways to reduce work-related risks.

Actions

With support from Exxon Mobil Biomedical Sciences, Inc., the Electrical Distribution Work Group analyzed the causes of its incidents and found a number of opportunities for Human Factors application.

Implementation

To reduce risks, numerous improvements were made to facilities, tools and work procedures to minimize manual handling, climbing and overexertion and to ensure that necessary information is readily available.

For example, portable test stands were installed in all 72 substations to eliminate the need to test breakers on line, reducing risks of electrical injuries and outages. Electric lifting devices installed in trucks and switchgear eliminated manual handling of breakers that could lead to back injuries. Proper power and hand tools reduced risk of overexertion.

Results

The work group did not have a recordable injury in the five years after the effort began. Unplanned outages were reduced 80 percent.

■ HUMAN FACTORS STANDARDS

A number of leading companies in North America and government departments are establishing standards for Human Factors in order to drive home the importance of this topic and at the same time improve their results. Most of the standards so far are in the area or ergonomics, that is worker/machine/ work environment interfaces. There has not been as much work done as yet on the consequences of management decisions and climate building. Because of its high importance, this must come in the future.

Some of the areas that Syncrude Canada, Ltd. has designed standards around are as follows: working environment, illumination, noise, work space access and egress, designing for handling and lifting, working postures, etc. An example of one of these is outlined, including references to government regulations.

■ CASE STUDY—NOISE STANDARDS

Syncrude Canada Ltd. Human Factors Standards

2.3 Noise

Occupational noise is the most common "on the job" health hazard. Exposure to high levels of noise can cause hearing loss. The nature and extent of the hearing loss depends upon the intensity and frequency of the noise and the duration of the exposure. Noise induced hearing loss may be temporary or permanent. Temporary hearing loss results from a short-term exposure to an elevated source of noise. After a rest period away from the source, normal hearing will return. If the exposures are

TABLE 9-2: NOISE LEVELS ASSOCIATED WITH VARIOUS LOCATIONS

Sound Pressure Level	dB
Jet engine at 25m (pain threshold)	140
Jet engine at 100m	125
Rock concert	110
Pneumatic chipper (jackhammer)	100
Heavy truck	90
Average street noise	85
Business office	65
Conversational speech	60
Living room	40
Library	35
Bedroom	25
Hearing threshold	0

repeated, permanent noise induced hearing loss may be the result. Noise-induced hearing loss is irreversible. It can be prevented through the use of engineering controls, administrative controls and the use of protective equipment. Studies of companies with an effective Hearing Conservation Program have shown reduced incident rates and lost time due to employee illness.

> A noisy environment...
> contributed to an operator's understanding that there was a need to "trip" a fan rather than a pump. When the fan was tripped, it led to a Coker going "offline".

2.3.1 Government Regulations and Syncrude Standards
The listed government regulations and Syncrude standards, as well as any others relevant to this topic, are to be met as a minimum.

2.3.1.1 Government Regulations
Additional information and requirements relating to this section can be found in:

> General Safety Regulation, Section 2 (Alta Reg. 448/83)
> Noise Regulation (Alta Reg. 314/81)

Recently proposed changes to the General Safety Regulation and Noise Regulation will require engineering controls to be implemented as the primary hazard control.

2.3.2 Syncrude Standards
Additional information and requirements relating to this section can be found in:

> Syncrude Specification A-9 (Noise Levels of Equipment)
> Syncrude Specification A-4 (Plant Noise Levels)
> ANSI S1.1 Acoustical Terminology
> ANSI S1.2 Method for the physical measurement of sound
> ANSI S1.4 Specifications for general purpose sound level meters
> ANSI S1.11 Octave, half-octave and third octave band filter sets
> ANSI S1.13 Methods for the measurement of sound pressure levels

TABLE 9–3: SUGGESTED MAXIMUM NOISE LEVELS WITHIN VARIOUS LOCATIONS

Workspace Area	Example	Suggested Limits	Voice Level and Distance
General Workspaces: areas requiring occasional telephone use or occasional direct communication at distances up to 1.5 m (5 ft.).	Maintenance workshops and shelters	75 dB	Raised voice at 2 feet. Very loud voice at 4 feet. Shouting at 8 feet.
Operational Areas: areas requiring frequent telephone use or direct communication at distances up to 1.5 m (5 ft.).	Operation centres, communication centres	Maximum 65 dB	Normal voice at 3 feet. Raised voice at 6 feet. Very loud voice at 12 feet.
Small Office Spaces, Control Centres: areas where communication must not be hindered.	Offices, command and control centres	Maximum 40 dB	Normal voice at 10 feet.

2.3.3 General Engineering Guidelines

Noise is a human factor consideration. A task requiring heightened levels of concentration or communication warrants noise levels below those requiring the use of hearing protection, (e.g., a panel man during an emergency shutdown). Table 9–3 provides designers with the suggested noise limits for various areas. *At the suggested limits, the table indicates the noise effects on communication (voice level and distance).*

2.3.4 Hazardous Sound Levels

2.3.4.1 Engineering and Administrative Controls

Engineering and administrative controls must be employed to reduce sound levels below or within the Occupational Exposure Limits (OELs) provided in Alberta Noise Regulation (AR 314/81)

2.3.4.2 Providing Personal Protection

If engineering and administrative controls fail to reduce sound levels to below or within the OELs, personal protective equipment shall be provided and a continuing effective hearing conservation program shall be administered.

2.3.4.3 Noise Exposure Levels

Occupational noise exposure levels shall be predicted, tested, monitored, and computed in accordance with the Noise Regulation (AR 314/81).

2.3.4.4 A Workers Maximum Daily Exposure

Occupational Exposure Limited (OELs) define a workers maximum daily exposure to noise without hearing protection. These limits are set out in Tables 9–4 and 9–5. The OELs take into account both the loudness of the noise and the duration of the worker's exposure.

2.3.4.5 Enclosures or Booths

Booths may be used to protect workers from noise exposure. Within booths lighting, heating and cooling must be sufficient to ensure comfort and to prevent a reduction in efficiency and work quality.

2.3.4.6 Location of Ventilation System

Areas used for control rooms, work-breaks, lunchrooms, conference rooms, training rooms, etc., should be located away from ventilation systems to mini-mize noise in these areas. This would include chilling units, furnace rooms' major supply and return vents. Should this not be practical, actions shall be taken to minimize the noise created by the ventilation system.

2.3.4.7 Noise Control Measures

Engineering noise control measures follow:

a. Substitution of machines:
   › larger, slower machines for smaller, faster ones;
   › step dies for single operation dies;
   › presses for hammers;
   › hydraulic for mechanical presses;
   › belt drives for gears;
   › spur gears are generally noisier than spiral or helical gears;
   › belt drives are generally quieter than gear drives; and
   › turbine drives are generally quieter than gear increases at high speeds.

TABLE 9–4: ALBERTA OCCUPATIONAL EXPOSURE LIMITS FOR NOISE

Sound Level (dBA)	Maximum Permitted Duration of Exposure Hours Per Day
80	16
85	8
90	4
95	2
100	1
105	1/2
110	1/4
115	1/8
Greater than 115	0

b. Maintenance of equipment:
   › replacement or adjustment of worn and loose or unbalanced parts or machines; and
   › lubrication of machines and use of cutting oils.

c. Substitution of processes and techniques:
   › compression for impact riveting;
   › welding for impact riveting; and
   › hot for cold working.

d. Dampening vibration in equipment:
   › increase mass;
   › increase stiffness;

TABLE 9–5: ALBERTA OCCUPATIONAL EXPOSURE LIMITS (IMPULSE NOISE)

Peak Sound Level (dBA)	Maximum Number of Impulses Permitted Per 8-hour Day
120	1000
130	1000
140	100
Greater than 140	0

> use rubber or plastic bumpers or cushions; and
> change size to change resonance frequency.

e. Reducing sound transmission through solid materials:
> flexible mounts;
> flexible sections in pipe runs;
> flexible shaft couplings;
> fabric sections in ducts; and
> resilient flooring.

f. Include noise specifications when ordering new equipment.

g. Reducing sound produced by air flows:
> intake and exhaust mufflers;
> fan blades designed to reduce turbulence;
> large low speed fans for small high speed fans; and
> sound pressure increases as a fifth of fan speed.

h. Isolating operators:
> provide a relatively sound proof booth for the operator or attendant of one or more machines.

i. Isolating noise sources:
> completely enclose individual machines;
> use baffles;
> confine high-noise machines to insulated rooms; and
> place sound absorbent material and covers around noisy equipment.

j. Flow of fluids in pipes:
> a Reynolds numbers less than 1000 should produce a quiet laminar flow; and
> a pipe size large enough to ensure a fluid velocity of less than 3.0 m/s or 10 ft/s should be used for water like fluids.

k. Mufflers:
> a dissipate muffler should not be used in highly contaminated air streams due to potential clogging and increased pressure drop; and
> reactive mufflers are only effective for their specific frequency range.

2.4 Vibration

The focus of this subsection is low frequency vibration (generally less than 100 Hz). Low frequency vibration is associated with whole-body vibration typically encountered in trucks, tractors, airplanes, etc.

TABLE 9-6: CHECKLIST FOR NOISE

Consider the task requirements and how the noise may affect the worker. Noise, like other subsections within Section 2, must be considered during the design of all new projects. The use of this checklist is meant to "direct" the user to questions regarding potential concerns surrounding noise. It is not meant to provide a "complete" list of potential concerns or questions.

No.	Questionnaire	Yes	No	NA	Remarks
1	Have the requirements of the General Safety Regulation: AR 314/81 and Noise Regulation AR 314/81 been met?	☐	☐	☐	
2	Have the requirements of the company's standards relating to noise been met?	☐	☐	☐	
3	Does the task involved require communication? If so, what actions can be taken to ensure noise levels are minimized?	☐	☐	☐	
4	Has an acoustical engineer reviewed the design of critical locations or sources of noise likely to be above 105 dB? Have recommendations been made as to how the noise levels can be reduced?	☐	☐	☐	
5	Will noise levels associated with this unit/equipment exceed 140 dB? If so, what actions will be taken so that workers are not exposed to this level of noise?	☐	☐	☐	
6	Will the background noise level in a control room likely exceed 40 dB?	☐	☐	☐	
7	Have vibration mounts or heavier materials been employed to reduce the transmission of vibration?	☐	☐	☐	
8	Are workers isolated or distanced from the noise sources?	☐	☐	☐	
9	Are in plant telephones stationed away from noisy locations or provided with a partial enclosure (both)?	☐	☐	☐	
10	Have exhausts been muffled?	☐	☐	☐	
11	Will noise from HVAC units effect personnel and tasks in adjacent rooms?	☐	☐	☐	
12	Are soundproof control rooms and lunchrooms required?	☐	☐	☐	
13	Will workers be able to hear alarms and communicate warnings?	☐	☐	☐	
14	Is lateral vibration in the range of 1 to 2 Hz reduced?	☐	☐	☐	
15	Is vertical vibration in the range of 5 to 16 Hz reduced?	☐	☐	☐	

Source: Based on Syncrude, Inc. documents.

Different frequencies affect different parts of the body. This is due to the following resonance frequencies of various body parts:

3–4 Hz	Resonance in cervical (neck) vertebrae
4 Hz	Peak resonance in lumbar (upper torso) vertebrae
5 Hz	Resonance in shoulder girdle
20–30 Hz	Resonance between head and shoulders
60–90 Hz	Resonance in eyeballs

Workers are most sensitive to vertical vibration between 5 and 16 Hz and to lateral vibration between 1 and 2 Hz. Note that women have been found to be more sensitive than men to vertical vibration above 3 Hz.

2.4.1 Seated Worker
Avoid or reduce the intensity of vibration in the 4–8 Hz range for seated workers as this causes the entire upper torso to resonate. Provide suspension seats.

2.4.2 Fatigue from Vertical Vibration
Minimize vertical vibration in the frequency range of 5 to 16 Hz.

2.4.3 Fatigue from Lateral Vibration
Minimize worker lateral vibration in the frequency range of 1 to 2 Hz.

Checklist for Noise
Consider the task requirements and how the noise may affect the worker. Noise, like other subsections within Section 2 must be considered during the design of all new projects. The use of this checklist is meant to "direct" the user to questions regarding potential concerns surrounding noise. It is not meant to provide a "complete" list of potential concerns or questions.

Case Studies

CHAPTER 10

Causes of Industrial Disasters and Lessons Learned

Past incidents (case studies) can be very useful when studied in detail to allow management and professionals to proactively reduce risks in their operation or project. From these studies you can learn about the basic causes, including their multi-factorial nature. These basic causes can often be sequential, producing a domino effect. You can also learn what actions were taken to reduce the risk after the incident. Although the particular incident may not be in your industry or location (country), it is often the case that the actions taken can heighten the awareness of some of the risks in your operation or project. This could be called proactive insight and can be very valuable to your risk reduction program.

Some of the basic causes might appear to be quite small and insignificant, if you were looking at it before the event. These past incidents can really drive home the message of how major damage to people, environment, assets and production can happen.

■ THE FLIXBOROUGH DISASTER
June 1, 1974

28 Casualties, 100 Injured

A chemical plant producing a nylon base chemical was located at the outskirts of a small rural village, Flixborough, located approximately 160 miles north of London, England. The Flixborough Plant of Nypro (UK) Limited was owned and controlled by two very large public corporations: Dutch State Mines and the National Coal Board of Britain.

First-class project management was demonstrated during the design and construction of Phase 2 of the plant.

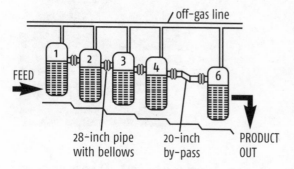

FIGURE 10–1: LAYOUT OF REACTORS AND
              TEMPORARY BY-PASS AT FLIXBOROUGH

off-gas line

FEED

28-inch pipe   20-inch   PRODUCT
with bellows   by-pass   OUT

The latest safety considerations were built into this plant. Phase 2 carried out the process of oxidizing cyclohexane to cyclohexanone, which is used to produce caprolactam—a base material for producing nylon. In the process, cyclohexane was kept at a pressure of 81.8 kg/cm^2 (approx. 9 bar) and at a temperature of 155°C. Large quantities were present at the plant (approx. 300,000 gallons). It is important to note that cyclohexane has a low flash point of –20°C.

The oxidation process of cyclohexane involved a batch of six reactors in series. These were constructed in a way that would allow gravity feed of cyclohexane from one reactor to the next. The reactors were connected with very short 28" diameter stainless steel pipes and stainless steel bellows (expansion joints), see Figure 10–1. After two years of reliable production, reactor 5 was taken out of service due to damage. A crack in this reactor had developed as a consequence of pouring water over the reactor because of a significant vapour

leak from a small connection. In order to resume production a 20" diameter stainless steel by-pass was designed, constructed and coupled from reactor 4 to 6. Because of the difference in height, it was built as a "dog leg". This was much longer than the regular connection between the other reactors. The plant went back to production and operated successfully for approximately two months before the disaster struck.

Summary of Event

The explosion happened on a Saturday afternoon at 4.53 p.m., June 1, 1974. A huge vapour cloud of cyclohexane that found a source of ignition caused it. This cyclohexane cloud was released and fell to the ground when the temporary by-pass failed at both bellows. The liquid cyclohexane from all the reactors then vapourized to the atmosphere.

The whole plant site was devastated. The explosion was so powerful that houses four miles away from the plant had structural damage and windows blown out. The subsequent fire was in the same order of magnitude as the largest fire during the London Blitz in World War II. Fire trucks responded from the nearest town of Scunthorpe within five minutes and within half an hour there was a total of 30 fire trucks. It took more than two days to bring the fire under control and they were still cooling down certain areas eight days later. During the first two days of the fire fighting operation there was panic about radioactive release. This caused a short period of anxiety amount the fire fighting teams. The radioactive source was found to be intact. It was part of a small device used by metallurgical engineers.

Total Losses

PEOPLE

> 28 people were killed—mostly operators of the plant. It took quite a number of days to recover all the bodies because of the site devastation.

> 100 people injured.

ENVIRONMENT

> There was extensive black smoke from the fire for a number of days. The official reports do not mention any environmental damage; however, the potential was there. This event happened in 1974. In today's "environmental climate" this event would have caused a major public out cry, media coverage and a government inquiry.

ASSETS

> 2,400 homes, shops and factories were affected and damaged in some way. The offsite damages extended to approx. 8 miles.

> The whole plant was totally destroyed with replacement cost set at approx. $180 million (1995 $). The plant was never rebuilt at this site.

> There was a large inventory of chemicals that was lost (66,000 gallons of naphtha; 11,000 gallons of toluene; 26,000 gallons of benzene; 2,000 tons of anhydrous ammonia). It was estimated that 433,000 gallons of flammable liquids were involved in the fire. This mixed inventory added fuel to the fire.

PRODUCTION

> Estimated production and export losses, until the industry could meet demands, were set at approx. $120 million.

OTHER

> Many lawsuits resulted and were settled over the next few years.

> Not only was Nypro's reputation tarnished but also the whole western world's chemical industries. The communities in the area around Flixborough lost all their confidence in Nypro and their management team. This is the reason why the plant was never rebuilt.

Immediate Causes

Because of the poor design of the by-pass, i.e., too long a pipe between the bellows and lack of support, the by-pass failed and fell to the ground. This allowed most of the liquid in the reactors to vapourize from the openings in reactor 4 and 6 and formed a huge vapour cloud of cyclohexane. The cloud floated over the plant site and found a source of ignition (possibly a furnace) and exploded with a force equivalent to 40 tons of TNT.

The by-pass was capable of rotating because it was connected to the bellows extending from reactor 4 and 6. It would do so when under high temperature the by-pass pipe and both bellows would expand causing a twisting motion putting excessive pressure on the bellows, which was above its design capability. This rotation of the by-pass caused the bellows to fail.

Basic Causes

> There was no mechanical or civil engineer with the correct experience on the site staff when the by-pass was designed. For the most part it was a "quick and dirty" design carried out by a young, less experienced process engineer. He was under great pressure from senior management to complete the design and construction in order to return production to normal.

> Senior management had previously advertised for additional engineers of different disciplines for this site. They must have realized there was a deficiency of expertise. The two owner companies were very large, they would certainly have had a large number of engineers with the right experience that could have been sent to the site quickly to carry out a proper design. They could also have brought in consultants or involved the original contractor. Certainly, management pressure for production and lack of foresight to supply the correct expertise was one of the most influential basic causes.

> Available manufacturer's literature on the bellows noted that only straight connections are safe. However, this was not followed mainly because of the problem with the height difference between reactor 4 and 6. This brings out the point that one must refer to all design standards and previous literature on equipment, before making changes. Also the information material must be kept current and up to date. In addition, the process engineer could very easily have contacted the company that built the plant and checked the design basis with them.

> The construction personnel who built the new by-pass (maintenance) should possibly have noticed this very non-standard design and questioned its reliability. The point here is that all personnel in any operation should continuously be searching for risks in their day-to-day work.

> The owners of the plant were Dutch State Mines and the National Coal Board of Britain whose basic area of expertise was within coal mining and distribution and not within manufacturing of chemicals. This could be a contributing factor to the reason why the owners and senior management did not recognize the severity of the Nypro plant being without a design engineer.

> Management was not paying attention to previous incidents leading up to this major disaster. The crack in reactor 5 was caused by deviations from standard practices. Cooling water was used to condense the cyclohexane vapour leak. The nitrate concentration in the water was sufficient enough to cause a nitrate-induced crack in the reactor. This was a poor practice and it should have alerted the senior management that the organization was weak and deficient in technical skills.

Additional Lessons

> All APEGGA professionals should know their own limitations of their particular individual discipline and **it is essential for engineers, geologists and geophysicists to know the limited of their knowledge and expertise.** Whenever in doubt, ask for help.

> A team consisting of different disciplines, such as chemical (process) engineers, mechanical engineers, metallurgists, chemists, maintenance personnel and perhaps a consultant from the company who built the plant should preferably handle projects involving design changes. Had the by-pass been designed with the care of the original plant there would have been much less likelihood of any problems. A large percentage of major incidents are initiated when **changes** are made to existing plants. These changes can quite often be minor in scope and therefore do not receive the required attention.

> The large inventory of chemicals on the site contributed to the fire and the extent of damages. Whenever possible the inventory of hazardous materials should be reduced and when storing hazardous materials, they should be placed in a way that minimizes any potential damage.

Flixborough Works of Nypro was a first-class plant with an excellent design, but this in itself does not lead to a safe and reliable operation. Management has to be totally committed to safety and risk management and include all employees in the program activities and practices.

At the end of the BBC documentary on the Flixborough disaster, the Bishop of Grimsby spoke at the remembrance church service for the 28 people who died:

Quality of Life:

There is an extra price to pay in the very quality of our lives. The truth is we cannot satisfy our increasingly sophisticated requirements and meet all the growing appetites of our consumer society and still hope to retain the green and pleasant fields of English rural society in undisturbed peace and security.

Is it not true that as in every discipline we break through barriers of knowledge and technology? We do by these very advances, which are to enrich life, make life yet more precarious.

Just one slip and then the abyss.

The Bishop of Grimsby,
June 1974

All APEGGA professionals should use the above quote as a key reference in their work.

■ ■ ■ ■ ■

The case studies presented in this section are major disasters. Minor incidents and near misses happen more frequently and if they are well documented they can provide useful information on the lessons learned. Alberta Labour has a library containing cases, including many where injury and death were involved. It is particularly important that minor incidents and near misses are investigated in any company or project. In most instances they could have turned into major disasters. APEGGA members can be key players in instigating these investigations. The APEGGA code of ethics is an excellent guide to your responsibilities.

When studying a past incident, make sure you use the official inquiry report. Be careful of using media information since the media can be misinformed and lack the technical expertise required to present the material.

The cases below are reproduced with courtesy from The Engineering Council, UK. Each case is referenced to the official report where a full description of the incident can be obtained.

■ PIPER ALPHA—FIRE ON NORTH SEA OIL PLATFORM JUNE 7, 1988

167 Casualties

Contributory Causes

(a) Breakdown in communications and Permit to Work system at shift changeover.
(b) The initial explosion put the main power supply and the control room out of action.
(c) Regulations did not require remote but potentially hazardous event to be assessed systematically.
(d) The safety policies and procedures were in place, the practice was deficient, i.e., frequency of emergency training.

Actions Taken Afterwards

(a) Regulatory authority transferred from Department of Energy to HS&E (Health, Safety and Environment).
(b) Requirement within the safety case for the setting of risk education goals.
(c) Safety case to demonstrate that the safety management system is effective, comprehensive, quality assured and auditable.
(d) Many design requirements, i.e., provisions of emergency shutdown valves, temporary safer refuges.

Senior management were too easily satisfied that the Permit to Work system was being operated correctly, relying on the absence of any feedback or problems indicating that all was well.

Source: Lord Cullen, "The public inquiry into the Piper Alpha disaster," London: HMSO, 1990. ISBN 010113102X

■ CHALLENGER SPACE SHUTTLE DISASTER—MISSION 51-L JANUARY 28, 1986

7 Casualties

Contributory Causes

(a) Faulty design of a seal unacceptably sensitive to a number of in-service factors.
(b) The sub-contractor's management supported its technical experts but NASA over-ruled them. The sub-contractor management did not pursue the issue any further in order to accommodate a major customer. NASA recommended the launch.

(c) NASA management structure permitted internal flight safety problems to by-pass key shuttle managers.

(d) Safety, Reliability and Quality Assurance workforce was reduced partly due to budget and time pressures.

Actions Taken Afterwards

(a) Design re-evaluation included tests over the full range of in-service conditions.

(b) Formal objective criteria adopted for accepting or rejecting identified risks.

(c) Safety, Reliability and Quality Assurance department strengthened and adopted a system for deviation documentation and resolution, which included trend analysis.

Neither Thiokol nor NASA responded adequately to internal warnings about the faulty seal design.

Source: W.P. Rogers, "Presidential commission on the space shuttle 'Challenger' incidence," US, GPO, 1986.

■ HYATT REGENCY, KANSAS CITY JULY 17, 1981

114 Casualties

Two suspended walkways inside the hotel collapsed.

Contributory Causes

(a) The design of the suspension connections did not comply with the relevant building code.

(b) The decision to change the design was made by telephone, but was not documented.

(c) The structural engineer's design drawings did not clearly assign design responsibility to the steel fabricator.

(d) The structural engineer did not review the drawings sent back by the steel fabricator carefully enough.

Actions Taken Afterwards

(a) Responsibilities between project team members regularized.

(b) Structural engineers reminded that they assume overall responsibility for their designs.

As indicated by their stamps, these shop drawings were reviewed by the contractors, structural engineer and architect.

Source: US Department of Commerce, National Bureau of Standards, "Investigation of the Kansas City Hyatt Regency Walkways Collapse," NSB Building Science Series 143, Library of Congress Catalog Card Number 81–600538

■ THE TITANIC April 14, 1912

Over 1500 Casualties
The "unsinkable" passenger liner sank on her maiden voyage when the vessel struck an iceberg in the Atlantic.

Contributory Causes

(a) The usual practice for liners in the vicinity of ice in clear waters was "to keep the course, to maintain the speed and to trust to a sharp look-out to enable them to avoid the danger".

(b) The ship's radio officer was catching up on a backlog of communications. Some outgoing messages from passengers took precedence over ice warnings.

(c) There were 2,208 people on board with lifeboats to accommodate only 1,178. Additionally many lifeboats were not totally filled.

Actions Taken Afterwards

(a) Lifeboat or lifeboat and pontoon raft provided for every single person.

(b) New rules for watertight bulkheads.

(c) All vessels with over 50 persons equipped with a wireless and emergency source of power.

(d) Rockets to be used only as distress signals.

(e) International conference instituted (SOLAS, Safety of Life at Sea, still in existence today).

It is hoped that the last has been heard of this practice [see contributory cause (a)] and that for the future it will be abandoned for what we now know to be more prudent and wiser measures.

CHAPTER 11

Case Studies of Canadian Industrial Incidences

■ INTRODUCTION

The Canadian experience with major incidents is very similar in frequency and extent as that of the rest of the Western world. There are two key differences, however, that should be brought out:

1. Canada's cold weather winters play a major part in many incidents. Many of our major plants are in the north, such as Ft. McMurray, Alberta, where winter conditions demand very different design and operational practices. Often the plants and facilities are designed in warmer climates, such as California, and the designers experience of cold weather operation is limited.

2. Canada is a very large country in physical size, with a much smaller population. Long distance communication, transportation, pipelines, power supplies, etc., adds a complex mix to any company's facilities. This can add additional risks if not fully understood. Sometimes a facility of a company in Ontario operates differently from one in Alberta or B.C. because of the differences in culture. These facilities can all belong to one company but the culture differences must be integrated to reduce risk.

Two major Canadian incidents, Lodgepole and Syncrude Coker fire, illustrate the different type of experiences Canadians have to contend with. There have been a number of smaller incidents in Canada over the years that indicate the same symptoms. These incidences tend to vary from company to company. It would be well worth the effort to research Canadian companies files and relearn the lessons of the past.

Over the last 20 years a number of significant negative incidents have occurred in Canadian industry. We have chose the Lodgepole blowout and the Syncrude Coker Fire as case studies to illustrate for our professional audience how large companies properly handle emergency situations.

The owner companies and the primary and subcontracting companies involved in these case are all first-class organizations. They certainly learned from the experiences of these incidents. These companies continue to work in a safe and reliable manner, providing a sound contribution to the Canadian economy.

■ LODGEPOLE

The Lodgepole exploration site was located 20 km west of the hamlet of Lodgepole and 130 km southwest of Edmonton. The owner company had hired a drilling contractor to perform the reservoir development at the particular site in question. A boom in the oil and gas industry in the province of Alberta in 1982 led to rapid growth and development and in turn a strain on the drilling contractors.

The drilling exploration for gas at this site was to be performed in a series of stages. The actual drilling operation commenced August 10, 1982, and the well was drilled to a depth of approximately 3000 meters. The installation of intermediate casing followed, prior to drilling into the expected productive formation. On October 15, 1982, the drilling crew initiated coring into the zone to produce samples for geological purposes. The first two cores were obtained with no apparent difficulties.

Summary of Events

When the crew was obtaining the third core on October 16, 1982, fluids and sour gas from the formation entered the wellbore, producing "kicks". "Kicks" occur when the pressure of the reservoir exceeds the static pressure of the drilling mud. For 16 hours after the detection of the "kicks", the crew fought, to no avail, to regain control of the well. On October 17, 1982, at 14:30, the well blew out of control. For 41 days of this period, all the effluent was on fire (major). Control was finally regained on December 23, 1982.

Losses

People

> 2 well cappers from a subcontractor died due to exposure to H_2S and a further 14 persons were hospitalized.

> 28 people evacuated, 4 families temporarily relocated.

> Estimated monetary loss due to lawsuits against the owner company and others was approximately $6 million.

The most severe hazard to people was exposure to hydrogen sulfide gas (H_2S). Low concentrations cause headache, eye irritation, sore throat, nasal irritation, nose bleeds in children, pain upon deep inhalation and some shortness of breath. High concentrations cause immediate unconsciousness, permanent brain damage or death if rescue is not immediate. The smell of H_2S

was detected as far away as Edmonton. Edmontonians and others who were close to the well site could perceive the "rotten smell" for 67 days. Many were very upset and somewhat scared of the consequences. The upstream oil industry lost a lot of credibility.

Environment

Losses to the environment have been divided into three categories: air, land and water.

AIR

The blowout emitted substances to the air that were harmful to humans, animals, vegetation and the aquatic habitat and there was a significant degradation of air quality over a large area during the blowout. Small animals and birds and a significant alteration to the habitat for various wildlife species incurred long-term effects.

LAND

The owner company clear-cut some 290 hectares to facilitate reforestation and to reduce fire risks. Approximately 39 hectares of soil requires stimulation of natural biodegradation for full rehabilitation over a number of years.

WATER

Ground water contamination was localized near the well but was not considered an immediate problem due to the isolation of the well. Aquatic life was not seriously affected, but the long-term effects are unknown.

Contamination of Zeta Creek and the Pembina River did happen and monitoring was required over several years to observe long-term effects.

Estimated damages to the environment were valued at $4 million.

Assets

There were losses to drilling rig, wellhead equipment and service equipment. Estimated monetary loss to assets was $8.5 million.

Production

On the production side there were losses in the form of decreased natural gas production. The event had negative effects on the operation of the entire company due to a strain on personnel. There was also additional loss in the form of additional wages to the company who finally got control of the well. Estimated monetary loss under this heading was $30 million.

Total monetary losses for the whole event came close to $50 million. Estimates for total monetary losses were arrived at from consultation with various industry sources, all amounts appearing in 1984 dollar value.

Immediate Causes

> *Failure of casing pressure monitoring equipment.* This prevented the drilling crew from recognizing the occurrence of the initial kick.
> *Failure of the degasser.* A degasser removes unwanted H_2S gas from mud, its failure caused the drill pipe to be exposed to H_2S.

The pipe became brittle and eventually failed.

> *Insufficient mud.* Improper mud density and the lack of sufficient supplies of the proper mud on hand at the site greatly affected the efforts to regain control of the well. Mud of the correct density would have allowed less H_2S gas to get through and contact the casing pipe.

> *Drill pipe separation.* Hydrogen embrittlement caused the pipe to separate. This failure caused the drill pipe to be blown out of the hole, which in turn caused the initial blowout.

> *Travelling block hook latch.* Failure of this safety device prevented crews from regaining control via "Top Kill" methods.

All these immediate causes prevented the drilling crew from understanding the severity of the situation and from applying the proper kick control method, which could have prevented the blowout.

Basic Causes

> The drilling plan and program was not sufficient. The owner company should have been prepared to encounter sour gas as an expected case scenario. That is, hydrogen-sulfide (H_2S)-resistant pipe should have been used and other precautions taken. It would appear that they either did not have an effective safety and risk management program, or did not ensure the sub-contractors practiced it. The prime contractor always has the main responsibility.

> The drilling crew was generally well trained and experienced. However, certain omissions and errors occurred in the drilling practice, such as relaxing of standards for core No. 2 and 3. Less time was taken to perform core No. 2 and 3. Training must be backed up with solid practices. Also there may have been too much push for production over all other priorities.

> The owner company drilling foreman and the primary contractor's supervisor had been awake and working for over 24 hours when the well blew out. When the situation was critical but not out of control, they failed to request help from experts in Drayton Valley and Calgary, who could have provided a fresh look and a better judgment. All personnel in these types of projects, in particular APEGGA professionals, must understand when to request help before it is too late.

> There seemed to be an improper system in place to learn from past mistakes. A well had blown out on an owner company site near Lodgepole five years earlier. Learnings from the earlier incident were apparently not applied to prevent the Lodgepole Blowout from happening. This reflected poorly on the owner company's incident recognition, investigation and analysis system.

> The procedures employed did not take into account the failure of equipment. Furthermore the drilling personnel were not

trained in maintaining the equipment.

› Emergency response was not as effective as it should have been.

Recommendations

The following recommendations are taken from the Energy Resources Conservation Board (ERCB; now called Alberta Energy and Utility Board, AEUB) Lodgepole blowout report.

The Cause of the Blowout and Adequacy of the Owner Company's Actions

› The following types of equipment should be examined for design, capacity and operational problems to ensure that they are adequate for worst case conditions:

  › degasser;
  › casing pressure instruments;
  › travelling block hook;
  › latch, Kelly hose (oil hose);
  › blowout preventers (BOP); and
  › equipment for anchoring drill pipe.

› The industry and ERCB should take any action necessary to ensure that drilling operations are carried out in a safe and reliable manner, particularly in the critical zone of sour wells. Special procedures should be developed, documented and used for operations in the critical zone. These would include detailed instructions in respect of tripping in, tripping out, coring, testing and other operations where particular care is required.

› Standard kick control procedures should be reviewed to determine whether they allow for situations where equipment failures or other unexpected events occur during control of operations.

› The adequacy of the current training programs for drilling personnel should be reviewed and in particular consideration should be given to emphasizing that they are effective in ensuring crew familiarity with kick recognition and control.

› Mud system design and operation should be reviewed with respect to density, system capacity, back-up supply, scavengers and impact of H_2S on mud and the ability to pre-treat and recondition it.

Well-Control Activities and the Adequacy of Actions

› Human Resources and Employment (HR&E) should give consideration to the development of an adequate, possibly compulsory, training program for workers who might encounter H_2S or other toxic gases in substantial amounts.

› The ERCB should consider how the experience and expertise needed to control a blowout would always be available when required.

Key Lessons

› In any operation it is extremely important to ensure that all employees (company and contractor) are following correct practices and have a well designed safety and risk management program in place.

> While on the site, professional engineers must continuously look for deviations and risks and bring them to the attention to the appropriate personnel.

> If they do not get a positive response to remedy the situation, they must take appropriate "whistle blowing" action.

> It is very important to understand that the prime contractor has the full responsibility for all the contractors and sub-contractors under their direction. Professional engineers who will work as/for prime contractors must realize this responsibility and act accordingly. In fact if they are on a site where the contractor is not complying with good practices, they may have to alert the proper authorities (i.e., the prime contractor or HR&E) to shut down the work.

■ SYNCRUDE COKER 8–2 FIRE AUGUST 15, 1984

Summary of Events

On August 15, 1984 approximately 2800 bbls of hydrocarbons were released and ignited when an 18" length, 6" diameter carbon steel pup piece in the slurry recycle line of Fluid Coker 8–2 ruptured. A pup piece is simply a piece of 6" diameter pipe approximately one-and-a-half to two-feet long cut out of a main line in order to clean the whole length of pipe during a shut down. A coker is a major upgrading process unit that is fed hot bitumen from the extraction process and thermally cracks this feedstock down to various lighter products. The two Fluid Cokers at Syncrude are at least twice as large as any cokers in the world.

This particular pup piece failed due to hot sulphidation corrosion and eventual thinning of the pipe wall. The fire burned out of control for approximately two hours and was extinguished after four hours. Damage was extensive to the Coker process area. These damages were in the range of $100 MM while equivalent production losses were estimated at several hundred millions. The repairs required approximately 121 days for completion and start up of operation.

The underlying immediate cause was the installation of the incorrect material into the alloy slurry recycle line in the 1979 unplanned winter shutdown. The line is made of 5% chrome steel. The pup piece that was put back in and welded was carbon steel. This could not take the very hot and scouring oil coke mixture over the years from 1979 to 1984. The pup piece was incompatible with the process and ultimately failed. The basic causes associated with the incident were: lack of material control, deficient change/risk management processes, poor procedures enforcing quality assurance, poor transition between owner/operator and maintenance contractor, less than effective mechanism for contractor selection, and miscommunication amongst all parties involved.

Risk analysis techniques were used to identify, classify and establish controls for future risks. The controls were the basis for the recommendations that include: establishing an effective material control program, improved risk management tactics, stringent quality assurance standards enforcement, preparedness for unplanned shutdowns, compatible contractor selection criteria and more effective communication.

Historically, Syncrude has had excellent performance in terms of low incident frequency rates and high levels of commitment towards improving safety and risk management. The recommendations resulting from this investigation are essential for achieving a higher level of performance.

Introduction

This report examines the details and losses surrounding the rupture of the slurry recycle line, and subsequent fire of Coker 8–2 on August 15, 1984 at Syncrude Canada. The immediate and basic causes associated with this incident are identified and a risk assessment was performed in view of these findings. Key recommendations were developed based upon the fault tree analysis and simplified risk assessment. An analysis of the petroleum industry performance in terms of safety and risk management was compiled and compared to Syncrude's operating practices in this crucial area.

Incident Background

JANUARY, 1979

> Syncrude had been operational for approximately six months and had experienced inefficient operation and upsets to Coker 8–2. Solids content in the feed was approximately 20% due to upsets in other processes within the plant (current solids content is 10–11%). The cyclones, in the scrubber portion of the coker, were unable to handle the high

level of solids in the feed and consequently released high solid carryover to the scrubber and to the slurry circuit. This caused the plugging in numerous process lines, including the slurry recycle line. There was a crash shutdown as a result of process upsets, very cold weather, first operation of a new coker and most likely some inexperience and design problems. Hence there was an opportunity to clean the recycle line. An outside maintenance contractor, just hired in the previous few months, was responsible for cleaning the line by the following steps:

> 18" length of the 6" diameter 5% *chrome* steel recycle line was removed by a cold cut procedure;
> the whole line was cleared by high pressure hot water/caustic wash;
> re-assembled slurry recycle line with 18" length of the 6" diameter *carbon* steel;
> weld completed for a chrome to chrome steel connection, not for a carbon to chrome connection;
> weld was tested for hardness; and
> pipe was insulated and coker was brought back on line.
* *Note:* A number of other changes and cleanings were carried out at the same time to improve operations. Approximately 2,000 contractor personnel were on the shutdown job.

AUGUST, 1984

On August 15, 1984, four-and-a-half years after the wrong material (pup piece) was put back in to the slurry line, there was normal operation of Coker 8–2 at a feed rate of 82,000 BPD. At 21:30 the *carbon* steel pup piece ruptured due to hot sulphidation corrosion and thinning of pipe wall, over the four-and-a-half-year period. The rupture occurred at the mid-point on the gravitational side where the wall thickness was 0.038" (original thickness was 0.28"). Hot recycle slurry at 374°C and bitumen feed were released at a rate of approximately 86 BPM. A source of ignition sparked the hydrocarbon release and subsequent fire that burned out of control for two hours. The final fire was extinguished at 01:20 and the all clear was sounded at 03:30.

Detailed Losses

People
> No fatal or life threatening injuries occurred.
> Principal reported injury was hearing loss—this was temporary.

Environment
> 2760 bbls of liquid hydrocarbon were released, but then burned or cleaned up.

Assets
> More than $100 MM in property damage was caused.

Production
> There was several hundred million dollars in loss revenue.

The fire caused extensive damage to the coker and surrounding process areas. The feed to the coker was immediately isolated by tripping of the feed surge drum pump. This significantly reduced the available fuel for the fire. Other attempts were made to isolate feed lines and the emergency response significantly minimized the losses associated with this incident. However, the fire resulted in a shutdown of 121 days to repair the damaged equipment and facilities (normal shutdown was approximately 34 days). In addition to the losses summarized above other indirect costs were incurred. In terms of other losses, employee morale dropped and increased workplace anxiety was experienced.

Causes

Immediate Causes

1979
> Cold cut piece was not properly secured and therefore was not available for replacement— 5% chrome steel.
> Incorrect material was installed— carbon steel.

1984
> Failure of pup piece due to wall thinning that was a result of hot sulphidation corrosion over a four-and-a-half-year period.
Note: The immediate causes that led to the incident were a direct result of a very new organization, a new maintenance contractor and a very cold weather shutdown with several design changes and considerable repair work. The company management were not fully familiar with conducting a

100 ■ INDUSTRIAL SAFETY AND RISK MANAGEMENT

cold weather shutdown and had not yet built up a solid relationship with the new contractor.

Basic Causes

1) Lack of material control:
   › lack of proper material identification, storage and tracking procedures;
   › poor housekeeping during shutdown; and
   › lay down area inadequate for crucial items.

   The pup piece should have been properly tagged and secured to be re-installed immediately after finishing the cleaning. In addition, the correct alloy (5% *chrome*) was not clearly labeled which led to the improper material being installed.

2) Lack of change/risk management:
   › failure to identify all the key hazards associated with the task and implementation of effective control procedures.

3) Lack of proper procedures enforcement and quality assurance:
   › safe work permit system not effective;
   › lack of emphasis for critical work and welding; and
   › inadequate mechanism for double checking standards of completed task.

   Syncrude personnel and the contractors failed to ensure that all tasks were completed to the specified standards. Quality assurance was not effectively integrated by all concerned into daily activities.

4) Premature transition of maintenance responsibilities from original construction contractor to maintenance contractor:
   › limited on site experience and expertise in the overall operation of the unique, world scale facility at Syncrude;
   › Syncrude personnel worked through the operational problems to gain experience.

   The expertise and knowledge of the original construction contractor would have been beneficial in helping Syncrude resolve the maintenance problems that they had over the preceding years. At the time of the shutdown, Syncrude personnel were reacting to emerging situations instead of taking a proactive approach to the critical issues of the shutdown. This reactionary attitude was partially a result of the changeover between maintenance contractors.

5) Limited investigation of maintenance contractor compatibility:
   › the maintenance contractor was on site for approximately one month before a crash shutdown in January 1979;
   › required by law to hire Alberta Union workers and therefore had little familiarity with selected work force;

> worked on cokers similar in function but much smaller in size; and

> unfamiliar with the extreme temperature difficulties experienced in Fort McMurray.

Due to the unique operation employed by Syncrude, they limited themselves to a narrow sector of qualified contractors. Their choice was based upon the experience that contractor gained with the fluid cokers at the Imperial Oil Sarnia operation.

6) Miscommunication between contractors and Syncrude:

> confusion due to unplanned shutdown;

> lack of leadership, experience and supervision, both contractor and Syncrude; and

> scope of contract and alignment of objectives was not achieved.

The contractor employees lacked the direction necessary to complete tasks required to Syncrude standards. The contractor was working in an environment of poor communication and management of risks. The end result was that the contract workers lacked guidance from top management.

The fundamental basic deficiencies identified above contributed to the wrong installation in 1979 and the following major incident in 1984.

Industry Analysis

With respect to the safety and risk management performance of the industry, the oil and gas sector has been better than most, third only to the trades and service sectors through 1982 to 1985. Since 1985, the number of incidents has been declining consistently, yet with increased production, indicating a more holistic approach by management. Industry has become aware that an improved safety program can be directly responsible for increased profits, as money not spent on an incident is 100% profit.

These performance trends are the result of industry culture, which is subject to many factors, some strengths, some weaknesses. One trend is the increasing reliance on contractors. The decrease in the number of permanent staff may reduce the total operating costs, but it also leads to less ownership and personal responsibility in the workplace. This effect is amplified if the management-worker-union relationships have suffered in the past. Union employees have typically been assigned tasks that permanent workers deemed beneath them, giving the union workers a lower workplace status and poorer facilities, in addition to the increased pressure of finishing the job on schedule. All these factors combine to form a poor working environment.

Another inherent weakness of the oil and gas sector is the reliance on labour intensive shutdowns. Plant work sites typically must perform critical tasks as quickly as possible, as the entire plant production often relies on single units. Winter shutdowns, though avoided if possible, make matters worse, as the cold can affect workers' motivation, and attention to detail.

Perhaps the greatest effect on performance in the oil and gas sector are the ever increasing technological improvements and the availability of information. Industry is becoming more familiar with empirical performance capabilities and is understanding the limitations of material properties. Additionally, computer control and improved instrumentation are becoming available. This can allow field tests to be done rather than extensive laboratory analyses (Texas Nuclear Analyzer).

These problematic safety factors have been more than offset by industry's drive for improvements. The existence of consistent government policies and industry quality code parameters narrow the margin of error (Alberta Quality Program, Alberta Boiler Safety Association). These standards are currently receiving greater emphasis from upper management as the owners' liability for accidents is increasing. This is clearly demonstrated by Syncrude's continually declining incident frequency rate. Industry has followed this trend and Syncrude is considered a top-class performer in the area of safety and risk management.

Risk Analysis

A fault tree analysis was performed to identify the key potential failures associated with this incident. The level of risk associated with each basic cause was classified based upon a generic risk assessment criteria from Exxon and Syncrude. A simplified risk assessment was conducted in order to associate a level of risk and prioritize the basic causes. Controls were then established

to mitigate this level of risk and formed the basis for the recommendations.

Recommendations

1) Effective Material Control
   › All reusable or critical material must be tagged, inventoried and secured. The tag should identify material type, intended use, and responsible employee. This would include restricted access to those who are solely responsible for the material in question. This is accomplished through the creation of a well documented (i.e., material access log) and controlled lay down area.
   › Implement a stringent materials identification system.
   › Adequate inventory of material, readily available, must be distributed, documented and tracked from a central location.

2) Risk Management
   › Establish a procedure to identify hazards associated with each task. This involves a simplified method, such as a checklist, that can be used by all personnel. From these identified hazards the appropriate controls can be implemented to reduce the overall level of risk.

3) Quality Assurance Standards
   › Quality assurance must become a required part of the safe work permit system. This includes

documentation of task progress, results, and tests performed. Establish a mandatory verification process to ensure that documentation is complete before closing the work permit.

4) Unplanned Shutdown Preparedness
   › Develop appropriate multi-disciplinary task teams to identify root issues associated with the shutdown. The team must proactively assess potential failure modes and develop key strategies to prevent further re-occurrence.
   › Formulate a list that addresses all tasks that should be undertaken during this down time. This list will form the basis for assignment of responsibility and identification of associated risks.
   › Provide immediate access to expertise required to resolve issues.

5) Contractor Selection Criteria
   › Stringent criteria for contractor selection including; evidence of safety and risk management program, employee training, reference checks, and previous job history. This will indicate the suitability and compatibility of the contractor with Syncrude's needs and standards.

6) Communication
   › Management must become directly involved in the emphasis of a safe and responsible work environment. This includes creating a culture of safety as the top priority by visibility and clear commitment. Management must never place blame and play a key role into the integration of contractors into Syncrude's safety and risk management culture. Attendance at safety meetings or visibility on site are significant indicators of management commitment and involvement.
   › For effective communication between Syncrude and the contractor a continuous verification process for alignment of views and objectives must be established. This would include pre and post task meetings that identify critical issues associated with the task.
   › Work towards establishing a long-term relationship by ensuring that all policies and practices are congruent with Syncrude's. This is accomplished by continuous training and updating of management skills.

Effectiveness of these recommendations can only be achieved through training and development of all workers by increasing their knowledge and understanding in these critical areas. Actions must be assigned for accountability and stewardship must provide the backbone for their completion. Involvement and support of these remedial solutions from all employees is essential, but management must take the lead hand in implementing these control strategies.

Conclusion

The implementation and enforcement of these recommendations would have significantly reduced the hazards and risks that combined to produce this incident. Responsibility and diligence towards continuously improving safety and risk management must be present. The costs associated with these recommendations would be insignificant when compared with the overall losses that were incurred. Also, there was an extremely high potential for loss of life had the circumstances of the rupture been any different. The effective and immediate response to this incident minimized the overall loss to people, environment, assets, and production (PEAP). Incident investigations are crucial for recognition of casual agents, deficient management systems, and to recommend remedial actions for preventative measures. These findings must be shared on an industry-wide basis to improve the overall performance in the area of safety and risk management.

■ REFERENCES

Alberta Occupational Health and Safety Lost-Time Claims and Claim Rates. 1986: pp. 6, 8, 19.

—. 1987: p, 5.

—. 1988: p. 6, 18.

—. 1989: p. 6.

—. 1991: p. 6.

—. 1992: p. 6.

—. 1993: p. 106.

Interviews

Frank Bajc, Plant Manager, Alberta Power, Personal Communication (was with Syncrude at the time of the fire).

Dana Bissoondatt, Research Consultant, Information Services, Alberta Occupational Health & Safety.

Larry Bolt, Inspector, Alberta Boiler Safety Association.

Dr. Ken Lau, Chief Boiler Inspector, Alberta Boiler Safety Association.

Wayne Schiewe, Staff Process Engineer, Syncrude Canada, Personal Communication.

Harold Senger, Incident Investigation Team Leader, Syncrude Investigation Report on Coker 8–2 Fire.

Specific Topics

Contractors

■ INTRODUCTION

Contractors are essential to any modern company, especially in this current climate of downsizing. This present climate has made companies more dependent on contractors of all types (i.e., experts in computer software and management systems, as well as construction groups). Companies may have up to three times the amount of contractor personnel on site than they have of their own personnel, particularly in certain maintenance situations. Therefore, it is important that companies not only direct their attention towards their own employees but also towards the contractors in implementing an industrial safety and risk management program. It is crucial that the contract personnel have all the training and expertise that the company requires them to have. Contractors are for the most part very specialized people with a great knowledge within a certain narrower field. Because they are knowledgeable in this field, it is sometimes assumed that they also know how to work safely at the site to which they are assigned. However, this should never be presumed and the company should never allow any employee or contracted worker to work without adequate safety and risk management orientation and training, in particularly to the site.

■ HISTORICAL BACKGROUND

Previously contractors were required to do more dirty and less appealing types of work that the company's regular employees deemed not to be their regular type of work. Often the contractors were put to work without any training or orientation nor any deep understanding of safety and risk management.

Managers of the company, contractor management and the workers themselves often ignored the need for personal protective equipment. The contractor personnel often worked long hours and were under pressure to "get in, get the job done and get out." Contractors often would take short cuts to complete the job faster without realizing the implications of such a practice.

Investigations of incidents involving contractors' employees were often not carried out. They were regarded as being of lesser importance than the company employees. They were often treated as "second class citizens". The contractors were sometimes told by the company not to report small incidents for fear of investigation. It was certainly not recognized that contractors could give a fresh opinion on many issues and can be an important source of information for the company's risk reduction programs.

Contractors were the "Achilles heel" of an operation or project. Although a company would pay significant money for their services, they were not getting full value because of a poor company/contractor relationship. There are many examples where this inefficient relationship was one of the basic causes of major incidents. On October 23, 1989, the Phillips 66 Polyethylene Plant exploded in the southern United States. In this incident a vapour cloud explosion occurred, killing 23 workers and injuring 282 others. There was property damage of $800 million and $850 million loss in production. Trucks and cars in the parting lot 800' away were crushed flat. This was the largest single insurance claim in the world up to that time. Details of this incident can be obtained from the document produced after the incident inquiry—"Phillips 66 Explosion and Fire"; a report to the President, US Department of Labor, April 1990.

■ CREATION OF AN EFFECTIVE RELATIONSHIP BETWEEN CONTRACTORS AND COMPANIES

The performance of a company is only as good as its weakest link may seem to be common sense. Therefore, it is important that companies do not ignore their contractors but commit to establish an effective relationship between the contractors and the company and incorporate this commitment into a policy statement. As well as protecting workers' lives and health, many companies have found an economic incentive to giving safety and risk management a high priority when working with contractors. Unsafe work sites are expensive in both direct and indirect costs. Direct costs include medical expenses, Workers' Compensation, operations reliability and insurance. Indirect costs include reduced productivity, increased administration costs and equipment/property/environment damages. Owner companies can exert significant influence upon the safety and risk management performance of their contractor. This is done right from the bidding stage through to actual worksite completion.

Best of Industry Approach

> *Treat contractors as you would your own employees.* The best companies treat contractors in a similar manner to their own workers. Conversely, the average company does not understand the value of treating contractors well.

TABLE 12–1: EXAMPLE OF A POLICY STATEMENT
OF COMPANY A

**Key Point Summary of the Policy for
Contractors–Company A**

- Contractor management is responsible for directing and coordinating their work.
- Contractors must comply with regulations, standards and policies.
- Contractor companies will provide personal protective equipment and ensure it is used.
- Contractors must report incidents, injuries and property damage.
- Contractors must perform site inspections complete with critiques and corrective actions.
- Contractors will be evaluated on their safety, health and environmental performance. This evaluation will be of:
  a) individuals;
  b) groups or trades; and
  c) the contracting company.

They do not appreciate the high-risk situation in which they place themselves.

> *Downsizing increases dependence on contractors.* The best companies realize that as companies downsize their dependence on contractors increases and so does their risk.

> *Strive for continuous improvement.* Most companies record their own incident frequency rates. The best companies carefully record, analyze, review and improve not only their own frequency rates, but their contractors rate as well.

> *Investigate all incidents.* The best companies encourage and perform first-class investigation of **all** incidents, especially to people,

environment, assets and production (PEAP).

> *Incorporate programs that will address potential problems.* The best companies have recognized the problems with the historical relationship between the companies and contractors. Many companies have visibly responded and incorporated programs that address the problem.

An example of the policy of a company that has a proactive safety and risk management program for contractors is shown in Table 12–1 for Company A.

■ ONGOING EVALUATION OF THE CONTRACTOR

The commitment to hire competent contractors must include ensuring that all contractors with current contracts, and all potential contractors, provide information to the company on their safety program and safety performance. The information can consist of the following:

> A written safety policy endorsed by the contractor's top management.

> A copy of the contractor's safety manuals.

> A description of the contractor's program to ensure the existing employees understand the safety policy and procedures.

> A description of the initial employee safety orientation program followed by the contractor.

> The contractor's safety record for each of the last three years. This must include the following categories: numbers of fatalities,

lost time cases, total lost days, medical aids and modified work cases. The frequency per 200,000 hours worked should be calculated for each of the categories.

> The contractor's WCB experience rating factor and the industry average rating factor.
> A description of the contractor's incident investigation procedures and the types of incidents that are investigated. Copies of the incident investigation forms to be used should be included.
> A description of how often safety meetings are conducted, who presents and attends the meetings and how the topics are selected.
> A description of how the contractor's safety programs apply to subcontractors and the method by which successful implementation and compliance with the programs will be assured.

The company must ensure that the contractors are actually following through on their safety and risk management program and that it is not just sitting on a shelf. The prime contractor is fully responsible for the activities, including health and safety issues, of the subcontractor and this must be monitored at all times by the owner. Valuable information can be obtained directly from the contract workers during informal discussions and safety meetings.

Contract Management for Owner Company

A high quality and detailed contract must be drawn up between the owner company and contractor, including all the responsibilities and possible penalties for the contractor. The owner company will make sure that a sound contract is put in place and that competent contractors are hired. The project manager (engineers, geologists, geophysicists) must ensure that the contractor personnel receive the training and orientation to work on the particular site. It is also their responsibility to ensure that these personnel are fully involved in the safety and risk management program and particularly understand their obligations. The project manager must also take the responsibility of ensuring that contractors are treated as if they were company employees, particularly since they are often involved in high-risk jobs.

The contractor's management on site should ensure the following activities are carried out effectively:

> Any contractor meeting involving all personnel where important aspects of critical activity ahead is gone over and discussed.
> Team planning meetings in work groupings—These could be once a day depending on the type of work.
> Toolbox talks—specific topics, based on results and change of the contractor mix.
> Contractor performance critique by superintendents/managers.
> Special activity review sessions—proactive.
> Post shutdown/activity critique workshop. The contractors are and important source of information. This information should be recorded for future use in similar types of work.

Contractor–Professional Employees

Engineers, geologists and geophysicists can also find themselves employed with contractor companies and have to be aware of their responsibilities as a prime contractor. For example, section 3.2 of the *Occupational Health and Safety Act of Alberta*) states: "The prime contractor for a worksite is (a) the contractor, employer or other person who enters into an agreement with the owner of the worksite to be the prime contractor, or (b) if no agreement has been made or if no agreement is in force, the owner of the worksite. Further in section 3.4, the Act states that "the prime contractor shall ensure, as far as it is reasonably practicable to do so, that this Act and the regulations are complied with in respect to the work site."

■ APPENDIX 12–1: CONSTRUCTION OWNERS ASSOCIATION OF ALBERTA

COAA Safety Committee Mandate and Objectives

COAA Vision Statement

"No one gets hurt in heavy industrial construction."

COAA Safety Committee Mandate

The Safety Committee members will work collaboratively to improve overall safety culture and performance in the Construction Industry by:

A combination of identifying, developing and supporting implementation of leading edge industry safety philosophies, practices and tools that will contribute to improving cultural and safety performance for Owners, Contractors and Labour groups.

Guiding Principles

> Work collaboratively with owner companies, contractor companies, labour providers, educational institutions, government and other industrial associations.
> Openly share information to maximize lessons learned from incidents.

Strategic Objectives

1. Monitor industry trends for potential safety implications and risks that are or may contribute to injuries, incidents and claims using information from the WCB, Alberta Human Resources and Employment and Construction Associations to identify focus areas.
2. Identify, prioritize, develop, evaluate, keep "evergreen" and support the implementation and use of appropriate tools, models and Best Practices to address critical safety and health issues.
3. Develop and implement effective measures to reduce the number of serious injuries to new, young or returning workers (Workers at Risk).
4. Proactively influence changes to Legislation, standards and processes that impact the construction industry in Alberta.
5. Maintain effective communications with stakeholders.
6. Measure, evaluate and report on program effectiveness and implement appropriate actions.

Strategic Objectives and Action Plans

1. Monitor industry trends for potential safety implications and risks that are or may contribute to injuries, incidents and claims using information from the WCB, Alberta Human Resources and Employment and Construction Associations to identify focus areas.

   Action Plan:
   › Review and develop strategies to address critical trends and issues identified at the March 2000 Alberta Safety Summit.
   › Review WCB injury analysis information quarterly to identify trends and areas requiring further focus.

2. Identify, prioritize, develop, evaluate, keep "evergreen" and support the implementation and use of appropriate tools, models and Best Practices to address critical safety and health issues.

   Action Plan:
   › Develop a standard template process to be used for the development of Best Practices.
   › Selection of Best Practice topics.
   › Determining the deliverables/mandate of the committee.
   › Membership requirements.
   › Communications.
   › Timing and work processes.
   › Best Practices Workshop (May)—selection and delivery of workshop presentations.
   › Current Best Practices being developed.
   › Workers at Risk.
   › Behavioural Based Safety.
   › Fitness for Work (Quality Workforce).
   › Enabling in the Workplace.
   › Hazardous Materials Codes of Practice—consolidate inventory and post on WEB.

3. Develop and implement effective measures to reduce the number of serious injuries to new, young or returning workers (Workers at Risk).

   Action Plan:
   › Working with the Apprenticeship Program to enhance the safety comprehension of apprentices.
   › Ensuring new workers are visible in the work force.
   › Support and build on the new worker awareness campaign (Job Safety Skills).
   › Develop a Safety Mentor process.

4. Proactively influence changes to Legislation, standards and processes that impact the construction industry in Alberta.

   Action Plan:
   › Being involved in activities identified in Alberta Human Resources and Employment Business Plan for development, revision of the OH&S Act, regulations, discussion papers, etc.
   › Setting out a specific work plan each January with Alberta Human Resources and Employment.
   › Promoting Partnerships and Certificates of Recognition (COR) within the COAA.
   › Participate in GSR Review and provide feedback on proposed revisions to the GSR and OH&S Act.

5. Maintain effective communications with stakeholders.

   Action Plan:
   › Hold regularly scheduled meetings of the Committee— schedule set annually.
   › Distribute meeting minutes to all stakeholders.
   › Keep COAA Board of Directors up to date monthly.
   › Review and maintain the COAA Web site.
   › Share lessons learned in regard significant safety incidents, risks, alerts and/or innovations in a timely manner.

6. Measure, evaluate and report on program effectiveness and implement appropriate actions.

   Action Plan:
   › Monitor and measure program effectiveness following implementation of new practices (i.e., CSTS Program—injury analysis and learning retention evaluations).
   › Develop recommendations for improvements to practices based on evaluations.
   › Identify and develop scope for "new" modules for the CSTS program (May, 2000) and provide input to ACSA.

Updated: June 28, 2002
Copyright © 1999–2001
SOURCE: http://www.coaa.ab.ca/safety/Safetyobjectives2000-2001.htm
Used with the permission of
the Construction Owners
Association of Alberta.

CHAPTER 13

Small Company Performance and Program

■ INTRODUCTION

Small companies (50 employees or less) make up a large percent of businesses in Canada. In Alberta small companies account for 95% of all businesses and employ more than one-third of the workforce, which is typical across Canada. These small companies sustain 44% of all work place injuries. According to Canadian worker's compensation board records, many of the health and safety problems encountered by them are related to:

› noise;
› dust;
› chemical exposure;
› construction injuries; and
› back injuries.

Small companies do not have the resources to employ full-time health and safety experts such as physicians, nurses, industrial hygienists and risk assessment experts. However, the development and operation of a "safety" program in a small company must be the responsibility of the manager or owner.

■ STRATEGIES FOR SMALL COMPANIES

Small companies should develop strategies to address safety and risk management. These companies should:

› Develop procedures and practices to eliminate or control potentially hazardous situations.
› Investigate and learn lessons from all safety and risk management incidents, particularly near misses.
› Put in place and sustain a simplified safety and risk management program that suits the needs and budget of a small company.

- Understand and ensure compliance with provincial and federal laws on health and safety.
- Establish procedures and training in this safety area.
- Join the local industry association to network with companies who can provide guidance and expertise. Examples are the construction, pipeline, mining and producer associations. Contact with these associations can be made through provincial departments in charge of workplace health and safety.
- Locate appropriate resources that will provide services at **minimum costs**, such as:
  - government agencies (For example, Alberta Workplace Health and Safety, publishes a set of guidelines to assist small business in getting started with an effective program "Occupational Health and Safety Manual for Small Businesses" Parts 1, 2 and 3);
  - employer groups;
  - safety associations, such as Canadian Society of Safety Engineering (CSSE), Industrial Accident Prevention Association (IAPA) and Canadian Centre for Occupational Health and Safety (CCOHS); and
  - part-time experienced people and consultants.

■ SOCIOLOGICAL STUDY ON SMALL BUSINESSES

In 1988 Joan M. Eakin, Ph.D., and Karen M. Semchud, Department of Community Health, carried out a study on "Occupational Health and Safety and Small Businesses". A key objective of the study was to determine the barriers to improved occupational health and safety performance in small companies. Fifty small businesses in the Calgary area were studied—all with 50 or less employees. The study makes the following important points:

- The small company owner's perception revealed a widespread "no problems here" response to safety problems.
- There was a tendency to discount and normalize health hazards at work.
- A large number of the small companies were rarely subject to inspection by Workplace Health and Safety inspectors.
- Inspections performed by the fire marshal were taken much more seriously—perhaps caused by the respect for authority figure.
- Many of the small company owners believed they did not have the moral authority to intervene in certain areas of their employees' health and safety. The relationship between owner and employee was often of a social nature, with the owner and employees being "buddies" or relatives. Furthermore, the owner often expected the employees to know how to work safely without any training or attention.

Some Characteristics of Small Businesses

Small businesses are different. It is important to understand just what these differences are and work towards recognizing where they exist within the organization and design the safety and risk management program accordingly. The following characteristics were identified in the study, along with some ideas on how to make improvements:

> Small businesses are more sensitive to fluctuations in the overall economy, which causes the need to lay off and take on employees (i.e., more prone to financial instability and seasonal variations).

> "Permanent" sub-contractors, who would work alongside the regular employees, would perform some parts of the work. The owners did not feel as responsible to the sub-contractors as to their own employees.

> One half of the businesses studied had employees who were members of the owner's family, which sometimes would affect the way health and safety issues were dealt with.

> Many small companies, especially in the service and construction sector, have their workers scattered over numerous sites. This does not allow for good influence and control.

> Employees tend to do all kinds of tasks even though they may be specialized in one particular field. Small companies often require workers, who are flexible and can perform several tasks. However, the safety aspects of each of the different tasks are frequently not understood.

> Over 50% of the owners perceived health and safety to be of limited significance. The owners had very little to say when asked what they were doing to promote health and safety.

> The owners were much more concerned with personal protection against hazards than trying to reduce the hazards. They did not question what happens when a worker does not wear the equipment.

> Over 50% of the owners believed that health and safety is essentially the employee's responsibility. They provide the protective equipment and the worker should carry on from there and know how to work safely.

> Small companies tend to see incidents as part of doing business—it is a matter of being lucky or unlucky.

> Resources, both financial and technical, are reasons put forward by the owner for not being able to improve the safety and risk management performance. This reasoning is invalid since there are resources available at minimum cost.

■ IDEAS FOR IMPROVEMENT

The following are ideas for how to improve the safety and risk management performance for small companies.

> New role taken on by Workplace Safety and Health inspectors—not

only performing inspections but also providing advice, guidance, coaching and inexpensive programs.

› Creation of stronger industry associations to influence the small companies in order to convince the owners and managers that good "safety" is good business.

› Additional provision of effective training programs and seminars, by Workplace Safety and Health departments at a reasonable cost for owners/managers. Subscription to monthly magazines and other helpful literature on the subject.

› Larger companies who hire small companies as contractors should strongly influence them in the area of health and safety. It will benefit both their businesses.

› Worker's Compensation Board rate incentive can be redesigned to encourage small businesses to conduct their work in a safer manner.

› Conduct round-table conferences between management, labour, government and educators in order to establish a cooperation between the parties. Out of these round-table conferences should come ideas and recommendations to improve the safety performance of the small companies.

■ OCCUPATIONAL HEALTH AND SAFETY MANUAL FOR SMALL BUSINESSES

Information regarding occupational health and safety for small business can be obtained from provincial sources at very low cost. Any owner who operates a small business can easily interpret the material and install a program to suit the needs of their small company. Workplace Safety and Health operations personnel are available to provide advice. The elements of the program are outlined. Compare this concise program for a small company with the program detailed in Chapter 5, which is for medium and large companies.

Small Company Program Elements

Program elements for small companies to consider in developing an industrial safety and risk management program, based on professional experience, include:

› inspections;
› housekeeping;
› materials;
› elements of a Health and Safety Program;
› employee protection;
› fire protection; and
› health and safety hazards analysis.

There are many more small companies in industry than the larger types such as Dow, Shell, and Syncrude. The larger companies, with so much capital money involved, see the need for top-class safety and risk management programs and results. The small companies, which normally are used as contractors by the large companies, do not have the same funds or perceived incentives to work diligently on industrial safety and risk management. Some are shining examples of first-class performance in this field, but most are not. The larger companies have a major part to play in ensuring these smaller companies, which they contract, work to the same standards as themselves. That is, "John Doe Pipefitter Ltd." working for, say, Dow Chemical Ltd., should by following Dow's rules and regulations become well versed in the best of industry practices.

Government departments in each province across the country should ensure that particular attention is paid to these small companies.

CHAPTER 14

Risk Communication and Public Emergency Response

Risk communication, community awareness and emergency response capabilities are key factors in maintaining public confidence in the ability of any company to provide a safe and reliable operation or project. The public has a number of key concerns, including:

> The health and personal safety, both short and long term.
> Personal property damage due to fire, explosion or vapour clouds.
> The fears of unknown risks from an operation or project close at hand.
> Damage to the environment, both short and long term.
> Their potential for other losses, such as property values, traffic congestion and neighbourhood esthetics.

A sound risk communication and emergency response capability is essential for safe operation and addressing the concerns of the public.

■ RISK COMMUNICATION AND COMMUNITY AWARENESS

In today's "right-to-know" climate it is important that management show a commitment to openness and community dialogue on risks from their operation or project. The industry must maintain a community outreach program to openly communicate information to the public, actively listen to their concerns and respond in a timely manner. V. Covello and F. Allen have developed the pamphlet: "Seven Cardinal Rules of Risk Communication" for the US Environmental Protection

Agency (USEPA) (1988). The rules are stated as follows:

1. Accept and involve the public as a legitimate partner.
2. Plan carefully and evaluate your efforts.
3. Listen to the public's specific concerns.
4. Be honest, frank and open.
5. Coordinate and collaborate with other credible sources
6. Meet the needs of the media.
7. Speak clearly and with compassion.

A good relationship with the community is vital to successfully developing and communicating a company's risk management plans. Worst case scenarios and how incidents can be prevented should be communicated to the public. This communication task should not be avoided. In establishing the relationship with the public the following should be taken into account:

> the status of the risk assessment work on the operation or project;
> the present level of the company's relationship with the community;
> the current expertise of the company's communication resources;
> the status of the communication to the company employees—the first audience is the company's own employees; and
> the state of the company's communication work plan and its implementation.

Management must set up and continually update a communication work plan, including designating key experienced speakers. These speakers must be well trained in the company's business and objectives. They must be prepared to talk to the public before, during and after significant changes and particularly in connection with emergencies. All employees should be provided with orientation and some training on the company's communication work plan. This is vital since they have important informal contact with the public outside of their workplace. The employees are the company's key ambassadors.

When preparing to communicate with the public, the company must:

> Ensure that the key designated and trained personnel are used and that the roles and responsibilities are delegated.
> Build on the existing credibility and relationships with the community resulting from previous communication efforts.
> Focus on the risk reduction actions the company has taken to prevent any public interference and/or damage.
> Concentrate on the work that has been done to prevent incidents and also key response capabilities in the event of an incident.
> Provide both oral and written communication to the public. Any written communication will be viewed as a public document.
> Be prepared to react to media coverage that will result from all these activities.

Each company, depending on their industry and particular project will obviously customize their risk communication work plan to suit their particular needs. Starting with the key points, specific details must be incorporated into any particular company plan.

Note: For further and in depth reading for this topic please refer to Canadian Standards Association.

■ EMERGENCY RESPONSE

Emergency planning and preparedness are essential to ensure that in the event of an incident all necessary actions are taken for the protection of the public, the environment and company personnel and assets.

Emergency response plans must be documented, be accessible and clearly communicated. These should not only be designed by the appropriate company personnel, but must have major input from and review by the community emergency services, i.e., fire, police and medical. Where there are other industrial complexes close by, it is important to become involved in mutual aid agreements. Examples of these can be found in almost every community in Canada where hazardous industries exist. A joint emergency plan involving the local industries and the communities work together to provide expertise and resources needed to combat specialized emergencies. For more information contact the Canadian Association of Fire Chiefs and refer to the Canadian Standards Association guidelines CAN/CSA Z–731–1995 "Emergency Planning for Industry"—a national standard of Canada.

Equipment, facilities and trained personnel needed for emergency response must be identified and readily available. These personnel can either be full-time members of the company's fire and rescue team or fully trained regular employees who respond to emergencies (or a mixture of both). On or off site medical personnel must also be involved and ready to respond. Simulations and drills must be scheduled at appropriate intervals to provide a state of readiness and ensure continuous improvement for first-class response.

Some of the specifics that should be included in the emergency response plan are:

> Details for combating any particular emergency that may occur in the company's facilities or projects.
> Evacuation plans for the public in the immediate surroundings in the event of a disaster. This should also include particular municipality areas in direct line with the disaster. Designated leaders must be identified and have a fully operational communication system at their disposal.
> Statutory requirements and communication responsibilities to the appropriate government departments and agencies.
> Detailed emergency procedures should be published and displayed in convenient and strategic locations, available for all appropriate personnel. Computer systems are very effective in this area, both for display and communication.
> A detailed list of available resources, trained responders and a full range of available equipment.
> Definition of roles and responsibilities.
> Refresher training of supervisors, fire fighters, rescue teams, medical

personnel and workers to ensure familiarity with how to effectively handle all types of emergencies.

› Systems available to broadcast any alarm or call for assistance, either in plant or nearby affiliated organizations. Familiarity with the locations of each facility and location and the routes that might be used.

The emergency response plan should also include a published list of company personnel who will be in direct charge of:

› combating any emergency;
› securing and protecting workers and equipment involved;
› safeguarding any equipment and other assets remaining after a disaster;

› investigating and reporting serious injuries to the correct authorities;
› covering transportation needs; and
› acting as liaison with the city, province and federal officials, insurance companies and news media.

Again in this case each company, depending on their industry and particular project, will obviously customize their emergency response plan to suit their particular needs. Most companies have a plan for the public response to emergencies and also a special plan for their own employees on their particular site.

Safety and Risk Management for Young Workers

■ INTRODUCTION

Since statistics have been kept over the last thirty-plus years, the Young Worker (15–24 years old) has had a much higher incident rate than the average industrial population. In 2001, Alberta statistics showed young workers represented 17% of the workforce (Statistics Canada Labour Force Survey) but accounted for 22% of the injured workers. These statistics are typical for all of Canada. This high incident rate of major injuries and death to these up and coming young workers is unacceptable and a safety and risk management program should actively address the common incidents that occur to young workers.

■ CAUSES

It is obvious that the young worker has less field experience than the average adult, particularly in the age group 15–24. Statistically, workers under the age of 25 are 33% more likely to be injured on the job than older workers. It takes considerable time working in the field—construction, chemical operations, oil refining, gas exploration, etc.—to gain avoidance of risk experience when working in the field.

Almost all companies, large and small, provide a safety orientation for the new worker. The new young worker often has trouble focusing on this orientation since they do not have a practical experience to imbed the information into their work habits. The average worker with experience will learn a lot more from these orientation programs and usually has had a number of them as they change from job to job.

Most companies emphasize that the supervisor or team leader should give special guidance to the young worker. This does not always happen, mostly because the supervisor/team leader is very busy and believes that the required guidance can be given at a later date, although this seldom happens.

Young workers tend to be in their first work experience. They do not like to ask questions of the supervisor/team leader in case they appear to be unsuitable for the job. This is summarized in the phrase "failure to ask when not sure".

The young worker observes the other workers closely as they are doing their work. Some of these experienced workers take short cuts, make mistakes and by-pass safety rules. The young worker tends to look at the situation and say to themselves "they must know what they are doing—so I should do the same." There is a negative peer pressure in some cases from the average worker's team. This pressure tends to say "why follow explicit management rules at all times?"

There is often a "macho" atmosphere in the workplace that says these accidents "can't happen to me". This is more prevalent in the case of the young worker since they have not experienced or witnessed many serious incidents.

In most companies there are people, systems and manuals where the young worker can find help in how to carry out a job safely. There is general tendency in this age bracket to either forget or know how to seek this knowledge.

Additional information on this topic can be obtained from Alberta Human Resources and Employment, Workplace Health & Safety. (The Web page is at www3.gov.ab.ca/hre/whs/ (May 2003); e-mail: whs@gov.ab.ca.)

■ TRAINING YOUNG WORKERS TO AVOID AT-RISK BEHAVIOUR

In an article in the *Alberta Occupational Health and Safety Magazine* (January, 2001), Ms. Marilyn Buchanan has a new perspective on this topic:

Why are young people more vulnerable to injury on the job? Most often, the actions young people take that endanger them are put down to lack of experience, lack of training or poor attitude. "They make bad choices," we say for everything from their driving to sexual behaviour.

The facts are not palatable: young workers are far more likely to be injured or killed on the job than older workers. In 1999, 24.6 per cent of Workers' Compensation Board—Alberta claimants were 24 years of age and younger.

Typically, our response is to shake our heads and ask, "Why don't they just smarten up?" New research shows that the answer may be quite simple...they literally may not have the brains for it.

Researchers are finding that young people's brains may not be wired for risk assessment. They simply do not have the ability to assign significance or importance to stimuli in their environment.

This means employers who hire young workers need to take a good look at their training programs for those in the 15- to 24-year-old age range. Unfortunately, many safety-training programs are designed for mature workers and are not effective when applied to younger members of the workforce.

Training must be designed to help young people develop their deductive reasoning and decision-making skills and their coping strategies.

Research Findings

Sandra Witelson, a neuroscientist at McMaster University, has found that during adolescence and early adulthood the brain's hardware is not completely connected. Recent MRI (magnetic resonance imaging) testing and other research show that the brain of a person between 16 and 24 still has a good deal of developing to do. In addition, the nerve pathway between the left and right sides of the brain is continuing to develop circuitry for comparative and abstract thinking.

We are born with all the brain cells we will ever have. What develops over time are the connections between those cells. Until these connections develop fully, a person is likely to engage in impulsive and risky behaviour.

While the connections are developing, the brain's electrical wiring does not transfer information effectively. This interferes with information feedback between the other parts of the brain and the prefrontal cortex.

The prefrontal cortex is one of the last parts of the brain to mature. It is responsible for self-control, judgment, emotional regulation and planning. This region of the brain is crucial to our ability to evaluate future consequences, weigh alternatives and select behaviours.

Dr. Scott Oddie, a neuroscientist and instructor at Red Deer College, suggests that the prefrontal brain region provides an executive function; it governs our ability to engage in rational, logical and analytical thought. Those approaching adulthood may not be able to appropriately evaluate the possible outcomes of their behaviour. They react more impulsively to the demands of their environment, without concern for the consequences. In addition, they lack experience, which can provide feedback on the basis of previous learning.

Adolescents are just beginning to develop the ability to envision different options and weigh alternative behaviours. They often perceive no relationship between themselves and their own actions. They overestimate their own abilities and tend to think, "It won't happen to me." In this age group, it is normal to be spontaneous and willing to take risks, and to feel invulnerable.

Learning to drive a car is a good example. Young people have the ability to master the basic skills required to control a vehicle, but their abstract thinking skills are underdeveloped, so they have difficulty assessing hazards and perceiving the possible consequences of their actions. And it seems that the development of these skills is not easy to speed up.

Helping Young Workers Stay Alive

In view of the latest research, it comes as no surprise that young people can not make all the measured, considered responses to situations that adults would like them to. So what can we do to assist young people during this high-risk stage of life? We have to help them to see risks as real and personal threats.

Research shows that the "rules and regulations" focus of many training programs for mature workers is not effective when used with young people. Nor are posters containing veiled threats. Fear-arousing messages can affect attitudes and intentions, but they have little effect on actual behaviour.

All stakeholders—employers, learning institutions, labour groups, associations and government, and not least, parents—can learn from the research and work together to train young workers to avoid at-risk behaviour.

You Can Train Young Workers To Make Good Decisions

If you are responsible for training workers under 24, keep in mind the traits (or peer group norms) that drive this group. Understanding their mindset will help you develop effective health and safety training, specific to their needs.

What Drives Young Workers?

Autonomy—This age group is working hard to act independently. Their behaviour is therefore often seen as rebellious. It is a necessary behaviour, however, for them to develop decision-making capabilities. An appropriate learning/education/training environment allows individuals to take some level of control for their learning.

Peer Influence—Also known as peer pressure. Can we turn this into a positive through positive modeling?

Risk Taking—Taking risks is often considered the "heroic" thing to do—a demonstration of autonomy or independence. This is particularly true of males. To grow, individuals have to be able to analyze and determine things for themselves, including identifying and managing hazardous situations. Working with young people requires that we acknowledge this tendency towards risk taking and find ways to manage the desire to test the limits.

High Audio/Visual Sensation Needs—The sensory/motor cortex and the vision and hearing centres are fairly well developed in this age group. But the connections between the emotions and thinking-brain areas are not working as well. Emotions tend to rule, resulting in the gravitation towards high visual/audio sensations. Hormones in the pleasure centres drive the need for fast-moving and loud activities.

Take Action!

The airline industry provides a good example of how to teach cognitive (thinking) skills. Student pilots take part in simulator exercises that allow them to experience possible piloting scenarios. Through this experience students learn to recognize inconsistencies, hazards, warning signs, etc., and how to manage these conditions. Training provided to young people needs to incorporate such learning opportunities.

CHAPTER 16

Provincial Regulatory Agencies and Workers' Compensation Boards

Workers' compensation is a no-fault workplace insurance system providing wage replacement, medical care and rehabilitation services to workers that experience an injury or contract an occupational disease. Although some nations may offer this protection to their workforce under their social assistance programs, many jurisdictions (particularly those in North America and Australia) provide this protection through an independent program that is directed exclusively at the workplace.

In these circumstances, the cost of insurance is borne solely by the employer rather than by the employees or on a shared basis, as might be the case in disability insurance programs. The direct benefit to employers who pay workers' compensation insurance premiums is that they are protected from potential financial ruin that oth-

erwise may be experienced in circumstances where employees suffer severe or fatal injuries. The rationale for employers solely bearing the cost of this insurance is that the cost of injuries or occupational disease incurred in the production or delivery of goods or services is reflected in the price. In order to remain competitive, employers are motivated to decrease or eliminate their losses in order to reduce the cost of their workers' compensation insurance. Ultimately, the reduction of workplace injury or disease also contributes to higher productivity, reliability and quality of goods or services, which further enhance the employer's competitive advantage.

In the U.S. and Canada, responsibility for workers' compensation insurance is decentralized and regulated by individual states and provinces.

Although similar in nature, the U.S. and Canadian approach in providing workers' compensation insurance can best be distinguished by the delivery model utilized.

A competitive model that has private sector insurers and insurers established by the State (State Funds) competing for employer business is primarily utilized in the U.S. In many cases, the state funds will fulfill the role of the insurer of last resort—providing workers' compensation insurance to employers that are unable to insure with a private sector carrier. Although this is the predominant delivery model for the U.S., a few states such as Washington and North Dakota elected to use a monopoly to provide workers' compensation insurance to their employer community.

In contrast to the U.S. model, the Canadian provinces and territories established monopolies as the sole providers of workers' compensation insurance. All work related injuries or occupational diseases for industries required to have workers' compensation insurance are insured by these monopolies, providing guaranteed benefits to the injured worker and no-fault liability protection to the employers. Some Canadian jurisdictions, such as British Columbia have included responsibility for regulating health and safety in the workplace as part of the insurer's role, while others (e.g., Alberta) have left the regulatory role with a government department such as Workplace Health & Safety.

The WCB-Alberta experience offers an excellent case study of a workers' compensation insurance program that is provided by a monopoly.

■ PROVINCIAL REGULATORY AGENCIES

Regulation of safety and occupational health in the workplace is primarily a provincial responsibility. In most cases the ministry responsible for labour or human resources and development are responsible provincially for occupational health and safety. There is federal government involvement that represents specific areas of responsibilities but for the majority of concerns the provinces have the responsibility. The Canadian Centre for Occupational Health and Safety (CCOHS) website (www.ccohs.ca) is a good reference for accessing information on various Canadian organizations of interest. (Although websites offer a wealth of information, be cautioned that it might not be the most up-to-date information, especially for legislation.)

Finding the proper information for a particular situation can be very complicated. Deciding on what is the appropriate regulation requires a careful search of the jurisdiction you are located in. In addition to provincial and federal legislation, there can be local municipal bylaws that can have an impact on a project. These apply mainly to zoning bylaws.

The Canadian Centre for Occupational Health and Safety has conducted a survey and point out the similarities in legislation across Canada. Many of the similarities concern the rights and responsibilities of the workers, employers and supervisors. Each jurisdiction manages each of these areas to meet their own need and often will include mandatory, discretionary or ministerial directed provisions in their regulations. According to the CCOHS study, each government is responsible for: enforcement of the legislation; workplace inspections; providing information; promotion of training; encouragement of new research; and a system for resolution of disputes. (A more detailed list can be found in Appendix 16–2: OH&S Legislation in Canada—Basic Responsibilities.)

■ WORKERS' COMPENSATION BOARDS

Workers' Compensation Boards are mutual insurance corporations with a monopoly in workers' compensation insurance that is regulated by various provincial and Workers' Compensation Board acts, for example the *Alberta Workers' Compensation Act*. Regulating occupational health and safety remains a provincial responsibility in Canada. (See Appendix 16–2 for a summation.)

History

The Workers' Compensation Board is charged with balancing the interests of workers and employers under a social contract put forth in 1913 by then Chief Justice Sir William Meredith to provide a no-fault disability insurance for the mutual benefit of workers and employers. The system, funded entirely by employers, provides an assurance to workers and their families that medical and rehabilitative care as well as indemnity for earnings loss is provided for. In return for funding the system, employers benefit from the protection from lawsuits that might otherwise arise. This has commonly been referred to as the "historical trade-off".

The principles contained in Sir William Meredith's 1913 Final Report (Ontario: King's Printer) are the foundation on which provincial workers' compensation systems are built. Since 1913, these principles have been enhanced to reflect changing times, but underlying concepts remain the same:

> Workers receive compensation benefits for work-related injuries at no cost.
> Employers bear the cost of compensation and in return receive protection from lawsuits arising from injuries.
> Negligence and fault for the cause of injury are not considerations.
> The system should be administered by a neutral agency having exclusive jurisdiction over all matters arising out of the enabling legislation.

Administering the Act

Each province administers their own Workers' Compensation Board according to their specific Act. The Workers' Compensation Board in Alberta is governed by a Board of Directors, appointed by government to represent the interests of employers, workers and the public and reports to the Minister responsible for the WCB-Alberta. Jurisdiction of the WCB-Alberta is limited to the province of Alberta, but coverage may be provided outside the province where the responsibilities of an Alberta worker require them to work on a temporary basis in a jurisdiction without comparable protection.

Worker Benefits and Responsibilities

The WCB-Alberta provides compensation benefits to any covered workers who incur injuries that arise out of and during the course of their employment.

Medical services provided to injured workers are the responsibility of the WCB-Alberta rather than the provincial health care program and may include such services as physiotherapy, chiropractic and rehabilitation. Although employers are responsible for paying workers their full salary on the day of the injury, the WCB-Alberta pays for their earnings loss (90% of their net earnings) for subsequent days, until the injured worker is capable of returning to work. Those workers who sustain a permanent disability as a result of their injury are eligible for a lump sum noneconomic loss in addition to their earnings loss payments, which reflect the clinical impairment remaining after recovery. In the event of a fatality, coverage provides for burial expenses as well as survivor benefits.

Although the WCB-Alberta will accept claims from workers that are filed within 12 months of the injury or illness, a workers' report of injury should be completed as soon as possible and forwarded to the WCB-Alberta. While on workers' compensation, workers should stay in contact with their employer and advise the WCB-Alberta of any change in their condition or of their return to work.

Employer Benefits and Responsibilities

The key benefit offered to employers covered by workers' compensation is the protection from lawsuits that would otherwise arise from workers being injured on their job sites. In a world that is becoming increasingly litigious, the value of this liability protection has significantly increased from when workers' compensation insurance

was first introduced. In return for the immunity from lawsuits, the cost of workers' compensation insurance is borne entirely by employers through premiums paid to the WCB-Alberta. This cost may not be passed on to workers.

The majority of employers in Alberta are required by law to carry workers' compensation insurance for all their workers. A few industries, such as farming and management consulting, are not required to carry this insurance, but may do so on a voluntary basis. A list of exempted industries can be found in the Workers' Compensation Act General Regulations or through the WCB-Alberta.

Workers employed in an industry covered on a compulsory basis are automatically covered by workers' compensation even though an employer may not have established an account. Employers who fail to pay premiums are subject to penalties and payment of premiums for the period in which the employer was operating without a WCB-Alberta account. Although workers' in a compulsory industry are automatically covered, employers are not. Whether in a compulsory or voluntary industry, employers (business owners, partners and directors of corporations) may choose to elect coverage (personal coverage) with the WCB-Alberta. With personal coverage, employers are eligible for disability benefits if they personally suffer a work related injury.

The WCB-Alberta collects premiums from employers to cover the costs of insuring work-related injuries. A premium rate is the amount paid per $100 of insurable earnings and is set annually. These rates reflect the risk of loss for individual industries. Those industries that contribute more to workers' compensation costs pay proportionately high premium rates than those industries that are safer. This industry premium rate may be further modified to reflect the performance of the individual employer. Those employers with poor accident records may pay as much as 80% more than the average premium rate for their industry. Exceptional performance is rewarded with a reduction of as much as 60% from the average for the industry. These financial incentives can impart a significant competitive advantage on exceptional performers over those employers with poor health and safety practices.

■ WCB-ALBERTA VISION

Every year over 36,000 Albertans suffer an injury that causes them to lose time from work. Tragically, each year sees almost 100 Albertans lose their lives to a work related accident, never to return to their families. This is a terrible human toll that also has a significant financial cost that requires Alberta employers to pay almost $400 million in premiums to the WCB-Alberta, not to mention the costs associated with productivity losses, equipment damage, product quality and reliability.

In light of the significant impact that workplace injuries have on individuals and the economy of Alberta, the WCB-Alberta has established a vision that looks to "Albertans working, a safe, healthy and strong Alberta." By encouraging employers to reduce injuries through financial incentives, awareness campaigns, research into best practices, partnerships with industry and labour, the WCB looks to make a significant contribution toward the reduction of this yearly toll. When injuries do occur, the WCB is there to provide for the security of benefits to injured workers, as well as apply best practices in disability management to assist workers in making a quick and safe return to work.

■ SUMMARY

Workers' Compensation Boards provide stability for the employee, the employers and the economy. The same fundamental principles developed by Sir William Meredith in 1913 for the Province of Ontario apply today in all provinces in Canada. The information in Appendix 16–3 from WCB-Alberta is typical of all WCB's in Canada.

■ APPENDIX 16–1: CANADIAN FEDERAL ACTS AROUND SAFETY AND HEALTH

The full text of many of the acts mentioned below can be accessed through the Consolidated Statutes Web page of the Department of Justice of Canada.

Part I – Acts

The Minister of Health has total or partial responsibility for the administration of the acts listed below and their related regulations.

Canada Health Act

This Act establishes the criteria and conditions which provincial health insurance plans or for extended health care services must meet to receive the full cash contributions under the Canada Health and Social Transfer.

Canadian Centre on Substance Abuse Act

This Act created an independent centre under the auspices of the Minister of Health to promote increased awareness of drug and alcohol abuse through a variety of information and educational programs.

Canadian Environmental Protection Act, 1999

The potential risks of environmental pollutants and toxic substances are evaluated under this Act that addresses pollution prevention and the protection of the environment and human health in order to contribute to sustainable development.

Canadian Institutes of Health Research Act

This Act establishes the Canadian Institutes of Health Research responsible for the creation of new knowledge and its translation into improved health for Canadians.

Controlled Drugs and Substances Act

This Act, passed on May 19, 1997, controls the import, production, export, distribution and possession of substances classified as narcotic and controlled substances.

Department of Health Act

This Act sets out the powers, duties and functions of the Minister which extend to all matters covering the promotion or preservation of the health of Canadians over which Parliament has jurisdiction.

Financial Administration Act

An Order adopted under this Act authorizes the Minister of Health to charge fees for processing drug submissions and establishes fees for providing dosimetry services.

Fitness and Amateur Sport Act

This Act provides for the authority of the Minister to enter into agreements with any province in respect of costs incurred by the province in undertaking programs designed to encourage, promote or develop fitness in Canada.

Food and Drugs Act

This Act applies to all food, drugs, cosmetics and medical devices sold in Canada, whether manufactured in Canada or imported. The Act and Regulations ensures the safety of and prevent deception in relation to foods, drugs, cosmetics and medical devices by governing their sale and advertisement and in addition sets out the labelling requirements for food.

Hazardous Materials Information Review Act

The Department acts as scientific and toxicology advisor to the screening agent of the Hazardous Material Information Review Commission responsible for reviewing material safety data sheet and label.

Hazardous Products Act

This Act controls the sale, advertising and importation of hazardous products used by consumers in the workplace that are not covered by other acts and listed as prohibited or restricted products. The Act covers consumer products which are poisonous, toxic, flammable, explosive, corrosive, infectious, oxidizing and reactive; workplace hazardous materials; products intended for domestic or personal use, gardening, sports or other recreational activities, for lifesaving or for children (i.e., toys, games and equipment) which pose or are likely to pose a hazard to public health and safety because of their design, construction or contents.

Patent Act

The provisions of this Act relating to patented medicine are administered by the Minister of Health effective in 1993. The patent protection for patented drugs is extended from seventeen years to twenty years. Subsequently, regulations were adopted linking the issuance of notices of compliance for generic drugs to the expiry of the patent protection period for the innovator drug. The Act establishes the Patented Medicine Prices Review Board and the mandate of the Board is to monitor and control the price of patented medicine.

Pest Control Products Act

This Act and Regulations is intended to protect people and the environment from risks posed by pesticides. Pesticides include a variety of products such as insecticides, herbicides and fungicides. Any pesticide imported, sold or used in Canada must first be registered under this Act which is administered by the Pest Management Regulatory Agency of Health Canada.

Pesticide Residue Compensation Act

The Act sets up a compensation regime under which the Minister of Health may compensate a farmer for losses suffered as a result of the presence of a pesticide residue in or on an agricultural product if certain conditions are met. One of these conditions is that the Minister issue a certificate confirming that an inspection made under the Food and Drugs Act disclosed the presence of a pesticide residue in excess of the permitted level prescribed under the Act, and that consequently the sale of the product would infringe that Act.

Quarantine Act

The Act authorizes the Minister of Health to establish quarantine stations and quarantine areas anywhere and to designate quarantine officers. These officers may inspect conveyances arriving in or departing from Canada, take protective measures against infested conveyances and their cargo and quarantine persons found infected with infectious or contagious diseases that would constitute a grave danger to public health in Canada.

Radiation Emitting Devices Act

This Act and Regulations prohibit the sale, lease and importation of radiation emitting devices that do not comply with the standards applicable thereto. The Minister of Health may appoint inspectors who are empowered to search premises and to seize and detain devices, and may appoint analysts to analyse or examine radiation emitting devices and packaging.

Tobacco Act

This Act establishes powers to regulate tobacco products, to limit youth access to tobacco products, to restrict the promotion of tobacco products and to increase health information on tobacco packages. It replaces the Tobacco Products Control Act and the Tobacco Sales to Young Persons Act.

■ APPENDIX 16−2: OH&S LEGISLATION IN CANADA— BASIC RESPONSIBILITIES

Are there any similarities in OH&S legislation across Canada?

Many basic elements (e.g., rights and responsibilities of workers, responsibilities of employers, supervisors, etc.) are similar in all the jurisdictions across Canada. However, the details of the OH&S legislation and how the laws are enforced vary from one jurisdiction to another. In addition, provisions in the regulations may be "mandatory", "discretionary" or "as directed by the Minister".

What are general responsibilities of governments?

General responsibilities of governments for occupational health and safety include:

> enforcement of occupational health and safety legislation;
> workplace inspections;
> dissemination of information;
> promotion of training, education and research;
> resolution of OH&S disputes.

What are the employees rights and responsibilities?

Employees responsibilities include the following:

> responsibility to work in compliance with OH&S acts and regulations;
> responsibility to use personal protective equipment and clothing as directed by the employer;
> responsibility to report workplace hazards and dangers;
> responsibility to work in a manner as required by the employer and use the prescribed safety equipment.

Employees have the following three basic rights:

> right to refuse unsafe work;
> right to participate in the workplace health and safety activities through Joint Health and Safety Committee (JHSC) or as a worker health and safety representative;
> right to know, or the right to be informed about, actual and potential dangers in the workplace

What are the supervisor's responsibilities?

As a supervisor, he or she:

> must ensure that workers use prescribed protective equipment devices;
> must advise workers of potential and actual hazards;
> must take every reasonable precaution in the circumstances for the protection of workers.

What are the employer's responsibilities?

An employer must:

> establish and maintain a joint health and safety committee, or cause workers to select at least one health and safety representative;
> take every reasonable precaution to ensure the workplace is safe;
> train employees about any potential hazards and in how to safely use, handle, store and dispose of hazardous substances and how to handle emergencies;
> supply personal protective equipment and ensure workers know how to use the equipment safely and properly;
> immediately report all critical injuries to the government department responsible for OH&S;
> appoint a competent supervisor who sets the standards for performance, and who ensures safe working conditions are always observed.

What does legislation say about forming health and safety committees?

Generally, legislation in different jurisdictions across Canada state that health and safety committees or joint health and safety committees:

> must be composed of one-half management and at least one-half labour representatives;
> must meet regularly—some jurisdictions require committee meetings at least once every three months while others require monthly meetings;
> must be co-chaired by one management chairperson and worker chairperson;
> employee representatives are elected or selected by the workers or their union.

More details about these committees are in the Health & Safety Committees Section on this site.

What is the role of health and safety committee?

The role of health and safety committees or joint health and safety committees include:

> act as an advisory body;
> identify hazards and obtain information about them;
> recommend corrective actions;
> assist in resolving work refusal cases;
> participate in accident investigations and workplace inspections;
> make recommendations to the management regarding actions required to resolve health and safety concerns.

What happens when there is a refusal for unsafe work?

An employee can refuse work if he/she believes that the situation is unsafe to either himself/herself or his/her co-workers. When a worker believes that a work refusal should be initiated, then

> the employee must report to his/her supervisor that he/she is refusing to work and state why he/she believes the situation is unsafe;
> the employee, supervisor, and a JHSC member or employee representative will investigate;
> the employee returns to work if the problem is resolved with mutual agreement;
> if the problem is not resolved, a government health and safety inspector is called;
> inspector investigates and gives decision in writing.

If you have specific concerns about what regulations require employers and workers to do, you should consult local authorities in your jurisdiction. This is especially true if your questions deal with the content, interpretation, compliance and enforcement of the legislation, and how it applies in your own workplace situation.

We have provided referrals in the section on OH&S agencies responsible for occupational health and safety. Local offices are usually listed in telephone directory "Blue Pages" or under separate federal and provincial government headings in other telephone directories.

Document updated on January 20, 1999.

Copyright ©1997–2002 Canadian Centre for Occupational Health & Safety

SOURCE: www.ccohs.ca/oshanswers/legisl/responsi.html
Used with permission.

Provides workers, employers and other groups with expertise and support to ensure safe and healthy workplaces.

What Is Workplace Health and Safety (WH&S)?

WH&S assists employers and employees in working together to meet health and safety standards. Consultations are made with employers and employees to help resolve health and safety issues, develop effective health and safety programs, and conduct health and safety inspections/investigations at work sites.

Alberta's occupational health and safety legislation is being updated in a three-step process. In the first step, the Occupational Health and Safety Act was amended in December 2002. One of the amendments permits the use of an Occupational Health and Safety (OHS) Code for detailed technical requirements and ensures the enforceability of the OHS Code.

In the second step, a new Occupational Health and Safety (OHS) Regulation took effect on March 31, 2003. The OHS Regulation deals primarily with administrative and policy issues.

The third step involves the release of the OHS Code. The OHS Code will contain detailed technical requirements in support of the OHS Act and OHS Regulation. Although its release is scheduled for September 2003, the OHS Code includes a six month phase-in period to allow employers to come into compliance with its requirements. This means that the requirements of the OHS Code are not scheduled to come into effect until April 2004.

Who is WH&S for?

The legislation covers approximately 107,000 employers in Alberta. Typically, WH&S serves employers, workers, worker representatives (union, medical professionals, etc.), health/safety associations, and health/safety professionals.

What services does WH&S offer?

WH&S provides occupational health and safety information to Albertans in a variety of ways including telephone inquiries, library services (reference and audio-visual materials), requests for technical information, and publications.

WH&S enforces the Occupational Health and Safety Act:

> Albertans can notify the WH&S of hazards at a work site. WH&S officers have the right to enter and inspect work sites to enforce the Act.

> An officer can order work stopped or call for corrective actions if the officer believes a work site is dangerous.

> WH&S receives notification of serious incidents and fatalities. Investigations are made of all work site serious incidents and fatalities. Reports are prepared on fatality investigations.

> WH&S can recommend a prosecution of an employer or worker if evidence reveals a significant failure to follow legislation.

> WH&S targets inspections for industries and employers with poor occupational health and safety performance.

> The Partnerships Program works with industry associations and large employers to audit and certify health and safety systems at work sites. This is a voluntary program.

Where can I access this service?

For inquiries, call toll-free anywhere in Alberta at 1–866–415–8690, between 8:15 am to 4:30 pm Monday to Friday. To report an accident, call toll-free from anywhere in Alberta at 1–866–415–8690, 24 hours, seven days a week.

Visit the Workplace Health and Safety website at www.whs.gov.ab.ca. Important Publications that are accessible at the website include:

An Employer's Guide to the Occupational Health and Safety Act (L1009).

Alberta's New Occupational Health and Safety Regulation—Highlights (LI021N).

Source: Reprinted with the permission of Workplace Health & Safety and Employment Standards Compliance and the Communications Branch, Alberta Human Resources and Employment Department.

CHAPTER 17

Key Benefits of Computer Systems and Communications

■ PROCESS CONTROLS

The key benefits of computerized process control systems are increased reliability, lower costs and greater accuracy. This greater accuracy means reduced control valve modulation that allows for control of process variables closer to their optimum values. For many processes, these optimum values are very close to critical process limit values where the process becomes unstable or upset. This ability to operate processes nearer to limits is also one of the pitfalls of computerization. The consequences of failures within automated control systems can result in losses of much greater severity. Computerized systems also tend to magnify the consequences of losses attributed to human factors issues (input errors, display design, alarm configuration).

The advent of supervisory control systems in the process industry has effected major change in the management control of plant and facilities. Typically in the less complex control systems of pre 1980 vintage, control room operating staff would rotate with field operating positions; this enhanced the communication and understanding of field and control related day-to-day problems. The increased complexity of "state of the art" control systems is making the rotation practice more difficult to administer as the skill sets of field and control room operators are growing further apart, in some cases engineers are being actively recruited for control room positions because of computer system skills. The pitfall is that the control room operators of the future may not have the thorough grounding in the process plant and the field exposure to ensure plant and equipment runs efficiently. This situation has to be addressed by training and orientation practices to avoid loss incidents caused by misunderstanding between field and control room operators. Modern systems make extensive use of simulators in the training programs of operators as often the simulated failures and changes in process variables will be the only way an operator can be trained to respond in an otherwise smooth operation.

Human factors consideration are many and varied in a modern control room; screen displays need careful colour planning to avoid read errors and eye fatigue. Layout, room and workstation size, audible signals, exercise facilities and lighting are all considerations to ensure operators are alert and able to respond to system upsets.

Alarm system management is an issue in very complex control systems. The Three Mile Island incident that occurred on March 28, 1979 in Harrisburg, Pennsylvania, provides a case study that demonstrates how an otherwise functional warning system can turn into an unmanageable volume of information and actually add to the severity of process failure.

On line by-passing of critical control parameters must be managed. The Chernobyl Nuclear incident that occurred on April 25th-26th, 1986, in the former Soviet Union, was a result of excessive safeguard by-passing to the point of system instability and the worst loss in recorded history.

■ MANAGEMENT OF SOFTWARE AND HARDWARE

With the increased reliance on control systems comes the added necessity of ensuring the accuracy of measurement devices and functionality of the control devices. Measurement devices can range from simple differential pressure measurement to radioactive source and detection devices. Control devices are typically valves but can also be variable speed motors and other more complex devices. Maintenance practices therefore must address the frequency of instrument checks and isolation of hazardous energy sources.

Sometimes it is necessary to use parallel or redundant devices with differing technologies on critical process variables. For example the liquid level on a high pressure vessel may have a flotation device, a differential pressure device and a nuclear source and detection device all sensing the level and Programmable Logic Controller receiving the input from these devices and determining the set point of the control valve. This adds to the demands of the maintenance team in terms of instrument check procedures.

Computer control systems also allow the added variable of application programs, for example the ratio between reflux and feedrate can be used to control other variables, say bottoms and side draw rates through a software application program to ensure optimum distillation column performance.

Change management practices become essential on these complex systems to avoid unauthorized change beyond pre-set limits and the resultant loss of production or impacts on product quality. Process safety issues such as system over pressure must also be considered when making changes.

■ STATISTICAL TOOLS AND ADMINISTRATIVE SYSTEMS FOR MANAGING RESULTS

With regard to the Safety and Risk Management Reporting and Recording systems, computer systems have enabled the creation of data base systems to simplify the process of analyzing and trending masses of information arising from risk management activities such as workplace inspections, incident reports, audit reports and findings and investigations. A typical computer based Risk Management Reporting System will have multiple data entry points, in some cases on-line completion of electronic forms. The system will be capable of producing trend reports and will be available for access by employees and management to access the records. Thus computer based statistical analysis is an invaluable tool in the communication of lessons learned from incidents to avoid recurrence.

Confidentiality issues must be addressed, for example, names of injured workers are usually not accessible by unauthorized personnel, also limited access authority "firewalls" must be created to ensure the safeguarding of information.

■ TYPICAL ADMINISTRATIVE SYSTEMS

While Incident data base management is the most important Computer Based administrative system there are several other applications that can help manage a multi faceted Total Loss Control Management System. These applications are limited to the experience of the author. As can be imagined the applications can be ever increasing as new hardware and software technologies become available. Fundamentally all applications store and manipulate extremely large masses of data, with modern connectivity techniques (local area networking and the Internet). This data and the knowledge gained from it can be made available to all workers within a company and potentially safety professionals in other companies.

Data base applications are the most prominent administrative tool, in addition to incident data bases these systems can track action item management (from investigation findings to audit reports), inspection tour actions, behaviour observation data and a multitude of activities where a large volume of data is gathered from field level risk management activities.

Work permit system management is increasingly being combined with equipment maintenance histories and even warehouse parts inventory systems in large integrated computer based applications. These systems facilitate reduced costs associated with spare parts storage and reduced downtime for equipment maintenance. They provide excellent "past history data" for modern Reliability Systems, which seek to ever increase plant and equipment "on-line" efficiencies or "up time." As a consequence, these systems can predict equipment failure and therefore prevent production loss equipment damage and the "domino effect" environmental emissions associated with complex plant start up and shutdown.

Computer applications are ever increasing in the areas of problem solving and risk based processes such as root cause analysis, fault tree analysis, hazard and operability studies, risk assessment and analysis techniques. Through these applications the vast quantity of actions, ideas and preventive measures can be tracked to completion.

As to the future, the hope is that the almost limitless ability for computer systems to store and communicate data will result in world-wide application development to ensure the safety professionals and managers of the future can learn the lessons from incidents happening outside their own companies, industries, countries and continents. In the future it will become possible to record an incident happening anywhere in the world and analyze it and make recommendations to prevent recurrence. The computer is a powerful tool in risk management, to enable it to be used to its full potential however, many cultural, legislative and legal barriers will have to be aligned to the vision of total loss prevention.

Building a Program

CHAPTER 18

Desired Safety and Risk Management Results Through Team Empowerment

■ HIGH PERFORMING TEAMS

High performing and empowered teams achieve the desired results of organizational safety and risk management. Within organizations, the human factor is effectively addressed through the implementation of empowered teams. Teams that are empowered value such characteristics as:

> appreciate the different strengths co-workers may possess;
> collaborative relationships between self and co-workers;
> common foundational values;
> established team norms and desired measurable behaviours;
> commitment to a common vision; and
> common processes necessary to build commitment and trust in their team.

They understand the value of common processes that denote how teams operate and cooperate. Some examples of common team processes are:

> clear roles, responsibilities and accountabilities of team members;
> effective problem solving and decision making tools;
> conflict resolution skills;
> planning process for team activities;
> clear boundaries;
> commitment to common goals and objectives;
> shared leadership within the team;
> empowering leadership and management through the different stages of team development;

> common mental models and concepts;
> communication in an open and positive environment; and
> understand the components necessary for building trust.

These operating norms create team synergy; the whole is greater than the sum of its parts. Many of us have experienced high performing team synergy, either through participation or observation. The human factor is addressed in the implementation of a teaming environment, and lessens the number of safety and risk incidents in organizations.

■ CHANGING ATTITUDES

The greatest challenge facing an organizational culture in transition is the creation of employee "want to" as well as "how to". So how do we achieve this when we know that we cannot change people's attitudes unless they choose to change?

What drives our behaviour and actions? How do we change our behaviours? Many training seminars address the modification of people and work skills within the confines of changing actions and behaviours. Often, however, following a "behaviour modification" program, when situations become stressful, people are prone to revert to previous "comfortable" behaviour.

It is only when we look below the surface of our own iceberg that we address our *self-skills*, which, in turn, create new behaviours and thus new norms. By looking to understand what, within ourselves, shapes our attitudes as well as how we engage people to want to change their attitudes, organization teams obtain the desirable

behaviours and actions towards safety and risk prevention management.

What Impacts Attitude?
Some key factors include:

> change without understanding how and what it will impact;
> a lack of understanding of the operating styles of employees;
> misalignment of our personal values within the organizational value;
> a lack of commitment toward or clear vision of the organization;
> unclear expectation and goals;
> unclear work and safety processes and systems;
> a lack of people processes such as decision-making, conflict resolution and effective communication;
> poor leadership and lack of management support;
> unclear definitions of roles, responsibilities and boundaries;
> too much confidence or confusion in the responsibilities of roles;
> a lack of job performance skills; and
> conflict between the employees and or leadership.

■ ORGANIZATIONAL EFFECTIVENESS

There are two features key to organizational effectiveness. One is desirable behaviour and the other desirable systems.

Management need to make sure both the systems and processes (work and people) are first in place in order to create the environment needed to practice safe behaviours. Management often understands the need for work processes, and address these in the workplace. However, the people processes are equally important. The tendency today is to look and drive toward bottom line results in order to create more profit for the shareholders; people processes are habitually ignored. When both the systems and processes are in place, organizations are able to focus on desirable behaviours.

The most effective and efficient way for organizations to meet their people process needs is through empowered teams. Leaders and management enhance teams through building commitment, achieving ownership (responsibility and accountability) and building trust. It is when these three criteria are in place that organizations are able to accomplish their desired outcomes.

■ ENHANCING TEAM RELATIONSHIPS

The four factors that shape our personality are: our core strengths and tendencies, innate abilities, environment and personal choice. The first two factors, our core strengths and tendencies as well as our innate abilities, are derived from our natural tendencies; these are stable and static. The last two factors, environment and personal choice, are dynamic and changing. Management impacts the environment and thus the culture of the organization through their systems and processes. Employees make personal choices depending on their values, beliefs and attitudes, needs and wants, personal vision and goals.

The first step toward empowered teams is the awareness of individual core tendencies and strengths, both of oneself as well as others. By understanding our core strengths as well as

our natural tendencies and operating styles, we can start to shape and build positive relationships between members of our team. We come into the world already possessing our core tendencies; they decide *how we move into action, how we make decisions and how we form relationships with our fellow workers*. When we understand out tendencies and natural operating styles, we can then learn to flex appropriately to styles that are not necessarily within the realm of our core tendencies, depending on the situation.

Thomas Concept™

One effective way of looking at natural tendencies or preferred operating styles is through the understanding and application of the Thomas Concept™. Jay Thomas* from Austin Texas, developed the Thomas Concept™ approximately 30 years ago. In studying leadership, he realized that two leaders with opposite strengths were still both effective leaders. Using Carl Jung's theory of personality types he developed a model that identifies the natural strengths and tendencies of people. It is an ideal model for organizational effectiveness since it focuses and addresses the people and relationship processes.

Its symbol is an infinity sign. On one side is our core strength, the natural way we tend to feel and act. The other is our supporting or flexing strength. We are able to utilize both sides of the model if we know how to flex to the opposite strength. By identifying our core, we are able to also recognize our supporting strength, thus

---

* Thomas Concepts Program: Jay and Tommy Thomas

appreciating why some actions are more challenging to us and therefore take more focus and energy.

This model also extends to the team; we can learn to appreciate the differences co-workers may possess. It helps us to understand what their natural way of being or operating styles may be, as well as the benefit of these strengths to ourselves as well as toward the team. It may also clarify how strengths or individuals can be effectively utilized within the organization.

The Thomas Concept™ is based on three pairings. These pairings include; thinking and risking; practical and theoretical; and independent and dependent (our dependent).

Some people are thinking (followed by action) while others are risking or action (followed by thought). Do you tend to think and plan what to do first or do you move into action and think as you are doing? Thinking strengths help us both in the development and implementation of procedures and policies. Those that enjoy this strength prefer to have a plan, are often more cautious, analytical and logical, may be emotionally reserved and often prefer to live their life with thought before action. People who possess risking or action strengths move into action quickly. They react quickly in any stressful or risky situation. Those who hold the risking strength tend to be more spontaneous, intuitive and are doers who prefer action followed by thought. No individual is a purity of thought or action. When given a challenging situation, both risking and thinking strengths can be effective. For example, on a climbing wall, the thinking strength will often step back and think about how to climb the wall effectively, while the risking strength will climb onto the rock face and plan as they move up the wall. Both are effective, just different. Both are **necessary** strengths in any organization.

Some people are more practical while others are more theoretical. Practical strengths have clear boundaries, drive to results, look after the "details", often judge behaviours, are able to deal with the facts and make decisions on what **is** rather than what **could be**. Theoretical strengths are "big picture" people who are able to look at many possibilities. They look at attitude, what could be rather than what is and brainstorm alternative outcomes. This pairing has the potential to cause the most conflict in teams.

The last pairing is dependent/out dependent or independent. Dependent strengths and operating style value other people's input into decision-making. They will naturally look for the best person to do the job and support them in their decision-making. They trust others and are able to rely on others. Independent strength will tend to "stand on their own" and make decision without the consultation of anyone else. They are most often self-motivated and self-reliant. They believe a job will be done well if they do it themselves.

When we explore, understand and appreciate the strengths and operating styles of others, and ourselves, we are able to work together as an effective team. When addressing shaping attitudes we need to appreciate and understand natural tendencies; the most appropriate strengths that come naturally can be utilized depending on the situation.

A high performing team utilizes all of the strengths that members of a team possess, either through flexing or by making sure the different strengths or operating styles are being utilized within the team.

Common Values

Values and guiding principles are another key area that is very important in achieving effective teams. The values and guiding principles, along with our natural tendencies or core strengths, form our beliefs and shape our attitudes around safety and risk management. These are our code of ethics and determine how we work with others.

We will be committed to an organizational direction only if our values and guiding principles are aligned with the values of the organization. As companies go through change, many employees can lose sense of direction with the belief that leadership values do not align with their own values. We lose commitment and buy-in, resulting in low team performance. This also directly impacts how safely we perform our functions and roles. Teams need to establish common values that unswervingly align with organizational values. Common values become the foundation for the establishment of operating norms and desired team member behaviours.

Values are the foundations that support the other four components of effective teams and organizations. *These components include: having a common vision; common goals and objectives (work and people); leadership and management; and common processes (work and people/relationships).*

Common Vision

A common team vision makes sure all members are on the same road, going in the same direction. The team vision must be in alignment with the organization's vision. Management needs to establish the organizational vision in consultation with employees. Buy-in is created when employees are involved and have input on direction. When employees believe in the vision, they are committed to the policies and procedures of the organization.

Goals and Objectives

Management needs to be clear on the organizational goals and objectives. The team needs to establish their own goals and objectives, addressing both the work or task goals as well as the people or relationship goals. These are the benchmarks that move us toward our vision. This enables employees to know both what they want to achieve as well as what their roles and responsibilities are in meeting their goals and objectives.

Common Processes

Common processes decide how we will work together. Every team is unique and therefore needs to define their individual processes. These needs are both work/ task and people/relationship related. Some examples of needs are: performance management processes; decision-making processes; communication processes; meeting processes; conflict resolution processes; and the definition of roles and responsibilities within the team. When processes are implemented, teams are able to define how well they perform tasks as well as how well individuals work with other members of the team and management.

Management and Leadership

Another area that needs to be addressed is the management and leadership within an organization. Chaos and confusion within the company directly relates to the work and people processes that are not in place. The people and work processes help to identify how we go about doing our jobs as well as the responsibilities within those roles. The role of management and leadership is to make sure the proper processes and systems are designed and ready for implementation. Fuzzy boundaries and unsure expectations result in lack of attention to the importance of policies and procedures.

The more .management and leaders can engage their employees in processes, the more often the occurrence of ownership and buy-in results. Employees understand when management effectively communicates what needs to be done. It is only when we involve others in decisions or direction that we engage not only the head (concepts) but also the heart (feel and believe) and feet (how to walk the talk). We then accomplish the commitment to decisions and direction. This creation of ownership causes people to pay attention to their role in maintaining and implementing proper safety measures.

Adults learn what they choose to learn. We cannot make them do anything that they choose not to do. Disciplinary actions and procedures often cause resentment and fear in the employees, thus creating safety and risk management issues. Operating this way may result in a change of behaviour but also results in poor attitudes toward jobs and leadership. Involving employees changes attitudes. Some leaders say this takes too much time. This time is well spent and in the long run cuts costs. Empowered people and effective teaming can only be accomplished through involvement. The organization achieves better bottom-line results as well as effective decision-making.

Role of Management and Leadership in Organization Change

Management and leaders also need to involve their employees in any changes within the organization. Increasing awareness of the reason for change in addition to input from employees will generate dedication to that change and therefore lessen the impact. The challenge for all employees and organizations is to build a positive attitude toward change. Change is difficult for some people. Patience and diligence is needed when implementing organizational change. According to William Bridges,* all organizations go through a "dip" in productivity when introducing major change. The varying transitions that employees experience depend on their core strengths. When management and organizations engage people, involvement occurs when change commences, thus easing the transition and lessening the impact on productivity.

Time will resolve the impact of change but may conclude with continued antagonism and anger; often some people will not let go. When communication transpires during change, the easier it is for people to adapt to that change. If something is going to impact us, we want to be a part of it. The more

---

* Managing Change and Transition: William Bridges

that we believe we are part of the changing direction, the less stress and more loyalty we will show toward our roles and responsibilities.

■ CONCLUSION

Employees should not be legislated. A teaming environment will enable employees to decide and have input into what affects them. We want to be involved and to believe we are valued. The role of management and leadership is to make this happen through the creation of desired values. Teams that engage in appropriate systems and processes, clear values and norms, a clear direction or vision, clear boundaries, clear goals and expectation and involvement of employees, will observe a change in the attitude of the people and thus as well success in reaching their bottom line (as defined by the organization). Benefits include increased enjoyment, commitment, achieve ownership through taking responsibility and accountability and trust building through sincerity, consistency and competency. Clear roles and positive attitudes will prevail in relation to the creating of a safe and low-risk workplace.

It is important to note that employees need to know the proper procedures and policies. These are the guidelines and boundaries. It is through the building of the "want to" that change in the attitude towards safety practices and tools will occur.

People want to be happy in their jobs. Employees want to feel valued. Any organization can be successful with the creation of a culture that builds commitment and trust through individual input of knowledge and ideas. Leaders and management must clearly define roles and responsibilities in addition to empowering employees through experience-shared leadership.

CHAPTER 19

Implementing an Industrial Safety and Risk Management Program

Although most companies have Industrial Safety and Risk Management programs in place, many industries in Canada and internationally will find this textbook very practical and useful for improving their existing programs. It can be used as a training program for new professionals, junior managers and supervisors already working in industry. Our Industry Advisory Committee critiques our program and material regularly, and ensures the use of the best updated material.

■ DEVELOPING A TRAINING PROGRAM

Developing a new training program should involve all levels of the organization. It is best to form a workable team from a cross section of the organization to spearhead the effort. This team can obtain ideas, workable content, etc., from all levels of employees. Then the team can assemble the program so that ownership and responsibility for **implementation** goes throughout the organization.

Very few companies have absolutely no Safety and Risk Management program in place. However, the program some companies have could be run-down, neglected and not very effective. It is our experience that this type of program should not be discarded entirely. There are always some effective parts and these should be used to build on. It is not usually the development of a program that is a major hurdle, it is the maintaining and continued practice of the various elements that become the problem. This is usually discovered after a major incident occurs, such as a life threatening situation, a fire or an explosion causing damage.

This textbook provides a valuable asset in designing a program, but as emphasized, it must be "bought into" by the whole organization in order that it is properly practiced.

■ THE BENEFITS OF A FIRST-CLASS PROGRAM

The key benefits of a first-class program are the reduction of risk to people, environment, assets and production (PEAP) for company personnel, contractors, public and investors.

When senior management realize the injuries, lives and loss of assets that can be save from a properly implemented program, they then put a major emphasis on a first-class, well-managed program. It sometimes takes a serious negative situation for management to realize this.

For example, in Alberta, approximately 120 deaths occur in industry each year. The total cost to industry annually is approximately two billion dollars. For all of Canada these statistics are almost ten-fold. Not paying attention to safety and risk management in the workplace can make the difference between staying in business or not staying in business.

Another key benefit, often overlooked, is the attraction and retention of first-class employees at all levels. Workers learn who are the "safe" companies to work for and apply to them. The first-class "safe" companies are almost always well managed and are good to work for, providing first-class long-term careers.

■ IMPORTANCE OF INDUSTRIAL RISK MANAGEMENT TO COMPANIES AND THEIR EMPLOYEES

In our experience, those companies who place major significance on risk management are the most successful in the total business effort. This is not to say safety and risk management is the only element to provide them with success but, without it, long-term success will not occur. Paying attention to people, safety, environmental protection, assets protection and continuous efficiency production makes a company successful.

■ IMPORTANCE OF SENIOR MANAGEMENT INVOLVEMENT

Company programs can fail if senior management do not back their safety and risk management program properly. The possibility of a major disaster occurring and continuous "safety" losses developing certainly are higher when safety and risk programs are managed poorly. Employees follow the lead of senior management and if senior management do not get involved and provide the necessary leadership, the whole organization will not take the program seriously, paying it "lip service" only.

When economic times are tough, some companies will go out of business by not paying attention to risk management. The blame almost always can be rooted in the lack of attention by senior management. They are so busy looking for ways to improve their operation that they neglect the importance of safety and risk management.

Almost all the successful Industrial Safety and Risk Management programs contain a major element on management leadership, commitment and accountability. Over the years, many companies have come to realize that this is the most important element of any program. Certainly management must provide the perspective, the climate and the goals and must allocate resources to ensure a successful and reliable operating program. Management also should ensure that the Industrial Safety and Risk Management program is integrated with all other company activities so that no matter what an employee is working at they would refer to and make use of the various elements of this program.

If management does not provide the backing, the program will fail or at the very best will produce poor results. The best of managers understand that safety and risk management provides a significant opportunity for managing costs and improving operational reliability. In the Afterward, Norm McIntyre, Executive Vice-President of Petro-Canada, provides an excellent example of senior management commitment.

■ CONCLUSION

Industrial Safety and Risk Management is based on over 30 years of field experience in industry throughout Canada and internationally by the principle authors and input from the University of Alberta's Safety and Risk Management program, founded in 1988 and developed over the years. During this period the Industry Advisory Committee, representing successful companies in the industry, have inputted their best of industry practices to provide the program with continuous improvement. The intention of this textbook is to provide a solid base for students and industry to implement, to manage and to improve their understanding and knowledge of Safety and Risk Management programs.

Afterword
NORM MCINTYRE
EXECUTIVE VICE-PRESIDENT,
PETRO CANADA

The following remarks are based on a keynote address Norm McIntyre, Executive Vice-President, Petro-Canada, delivered at the Process Safety and Loss Management Conference in Calgary, November 6, 2000.

Thanks also to all the organizers of the Process Safety and Loss Management Conference for this welcome opportunity to discuss loss management experiences. At Petro-Canada, I can definitely say that risk management is an important part of our business strategy and a measurable contributor to our profitability. We live in a dramatically more competitive business world than we did even a decade ago. It is a world in which every business leader is seeking

every possible advantage to drive profitability.

I doubt I can describe very much new on this topic given the level of expertise in this room, but perhaps I can share a senior management perspective on the topic.

Unfortunately, this competitive dynamic has not always worked in favour of improved process safety and loss management. This is because loss management must, quite reasonably, compete with other corporate priorities for scarce resources. Where loss management all too frequently loses out is that it is not viewed as a direct contributor to the bottom line. In the struggle to push dollars to the bottom of the balance sheet, loss management is too often seen as a cost to be controlled.

In ten years of meeting with analysts to discuss business strategies and corporate performance I have never been asked about our loss management systems. Rather, I have sometimes seen companies with high and escalating incident rates given excellent market assessments.

Loss management has an image problem. But I think it is fair to suggest that its profile needs to be raised in many organizations. To begin this process, loss management professionals must elevate their discipline to the level of strategic importance within their organizations. To do this, they must speak and act strategically in order to gain executive understanding that loss management is actually an indispensable component of effective business management.

When you get that message across, you begin to integrate loss management into the organization and you begin the move toward loss management leadership.

When that happens, loss management is no longer a policing game. It becomes a contributor to the bottom line and an important performance indicator. Loss management becomes part of the vision, decision-making, planning and benchmarking of the company.

Loss management goes beyond the prevention of loss incidents. It prevents losses that result from poor planning and design and creates a higher performance potential, through improved reliability and lower operating costs. The organization achieves an increased level of competitive fitness, enabling it to perform—and deliver—at a higher level in whatever economic conditions prevail.

■ PRESSURES

The pressures on business today are clearly greater than they were a decade ago.

Free trade has led to a global competition for markets and capital. There are no secure markets and every day's business headlines record another massive merger aimed at improving the competitive positions of the participants. In the oil industry, we are certainly aware of this phenomenon.

At the same time, the advent of ever-larger mutual funds has developed huge capital pools.

On the surface, this would seem to be an advantage to business in general. But we know it has not worked out quite that way. Capital has tended to focus on hot sectors of the economy, the leading example being electronic technologies. Process and resource extraction industries have been notably less in vogue. Today, we find ourselves competing all the more fiercely for capital while the huge money funds demand ever higher returns on investment.

Many of us would agree that these trends have contributed to improved economic performance worldwide. But we might also agree that they have promoted a certain amount of quarter-to-quarter, short-term thinking

In this environment, cost control is king and this can often lead to reduced investment in many areas, including loss management.

Meanwhile, performance standards and the costs of good corporate citizenship rise daily, in step with emerging new concerns for the environment, health and safety. Put simply, the demands of loss management are rising at a time when business pressures leave corporate executives with less room to maneuver.

If loss management is not properly positioned and promoted within our companies, it can be very tightly caught in a two-way squeeze.

I see this as a dilemma that can serve as a catalyst for change.

■ OUR EXPERIENCE

At Petro-Canada, events in the late 1980s and early 1990s provided our loss management specialists and their leaders with the opportunity to make their case for change.

We faced a rash of fires, spills, mechanical breakdowns, accidents and other losses in both upstream and downstream operations.

There was loss of production, asset damage and even loss of life. Insurance costs rose. Reliability was low and the cost of production was rising. We have first-quartile performance objectives that were being thwarted by losses and down time fairly consistently.

Clearly things had to change and I am happy to say they have. Over time, we have moved to integrate management of the environment, health and safety, reliability and physical risk into a Total Loss Management system.

When I say this is an integrated system, I have two meanings in mind. First, that loss management is not a policing action imposed by an isolated department of specialists. Rather, we have worked to make it a major part of "the way we do things around here"— something that is a part of everyone's job description at every level. Second, risk assessment and loss management are an integral part of normal business planning and operations. They are part

of our vision, our leadership model and our culture.

The result is that loss management is now seen as a front line tool for delivering on performance objectives. It is understood that you cannot consistently deliver first-quartile performance without first-quartile loss management systems and execution.

We have not achieved loss management perfection. There is still more opportunity in the system. But we have moved fairly far along the continuum. I would like to identify some of the landmarks along the way and discuss what others might learn from them, whether or not they have experienced similar adverse situations.

■ STRATEGIC POSITIONING

At some point, the logic of loss management can become very simple and compelling. That is because the consequence of inadequate loss management is increased risk—which sooner or later results in a loss event.

The immediate impacts can be severe. In the worst case they may include loss of life. Others are:

> lost human productivity;
> environmental damage;
> lost assets;
> lost production;
> replacement costs of assets and production; and,
> rising insurance costs.

I should say at this point that inaction by a company or an industry, or lack of successful action, invites government regulation or legislation. Typically these imposed solutions are much more costly and potentially less successful than a well-designed voluntary initiative.

This was certainly our experience at our Edmonton refinery. When the government mandated a resolution to high pH levels in water discharged from our refinery, the remedy cost $1.4 million. We later calculated that a voluntary solution might well have saved about $1 million of this expense. It was another of the learning experiences that motivated us toward proactive loss management.

Equally important, loss management controls and strategies affect reliability. Many solutions are a matter of improving management processes and these carry little or no capital cost burden.

Against these very modest investments we see very big potential paybacks in terms of reduced down time. We traditionally think of safety indicators as the primary measures of loss management and that is not necessarily a bad bias. But it is a bias—it is worth remembering that many companies would display a loss profile similar to Petro-Canada's. In our case, safety and fire losses have accounted for only 6.5 percent of total losses experienced by the business over the past 10 years. Reliability and other losses constitute a much larger and frequently underappreciated opportunity for business improvement.

While immediate impacts of loss events are cause enough to justify preventive action, the long-term, strategic impacts of inadequate loss management are equally compelling. They include damage to:

> reliability;
> revenue/profit;
> EHS performance;
> the quality of products and services; and,
> reputation and brand image.

The last item is bigger in a business sense than many people are willing to recognize. It can affect your ability to hire the best people, your ability to form strategic partnerships and your sales.

It is not that you get rewarded for doing this stuff right because, most often, people do not notice. The point is that they very definitely do notice when you do it wrong and they will punish you accordingly. In our industry, one need only mention the name of Exxon Valdez to make this point.

As I mentioned earlier, no one asked in advance about your corporate loss management system. But in the event of a serious incident, very pointed questions will be asked after the fact. When that happens, publicly traded companies frequently experience adverse impacts on their ability to raise capital and any company in that situation can expect its new projects to encounter diminished public and regulatory support.

I can think of one instance in which a very respected company had a plant expansion project delayed for more than a year when residents complained to the regulator about plant reliability issues.

■ STRATEGIC ACTION

Viewed in this light, loss management is a strategic initiative that must receive strategic attention.

I stress strategic action, because it is very possible—and it frequently happens—that loss incidents inspire self-defeating, nonstrategic action.

That is because tolerance in North American society for risk is moving toward zero, as public reaction to major events demonstrates.

This frequently leads to a shotgun approach to risk management. It is a simple concept—Blast all risks as they rear their ugly heads.

But this approach is potentially— one might say inevitably—expensive. And cost/benefit analysis is likely to be negative. Very few risks can be reduced to zero without eliminating the underlying business activity.

Loss management that is credible, sustainable and strategic must quantify and rank risks. In this way, the right issues are addressed in the right order to achieve greatest and earliest reduction in total risk profile.

Another favorite nonstrategic approach is—set a target and then shout at everyone continuously. This one looks strategic because there is a quantifiable objective. But if there is no management system or cultural support for the goal, it merely becomes a club with which to beat the next line department that encounters a serious loss. The same logic is captured in that wonderful old saying, "Floggings will continue until morale improves."

There has to be a better way—and there is. It starts by linking loss management to the business strategies and goals of the company.

A recent study shows that top performing companies are at least average in their adherence to three value disciplines and that they make one of these a passion, a part of who they are and how they do business. The three areas are:

> operational excellence;
> customer relations; and,

> product leadership.

For companies involved in resource extraction and processing, it would seem that operational excellence is the choice that can provide the greatest potential gain. It is certainly the one favoured by Petro-Canada and many of its peers.

In the production of commodities— such as oil, gas and petroleum products—one of the keys to operational excellence is a low-cost operating strategy. Loss management protects the low-cost operating strategy. Of course, loss management protects the low-cost position by keeping costs of accidents, process interruptions and physical losses to a minimum.

Having said that, I would expect most people here to be thinking that this last point is pretty obvious. I would have to agree. It is so obvious that it is taken for granted. It is rarely made explicit.

My point is that, where loss management is not made an explicit component of strategy, it never receives the attention or the dedicated execution that it requires.

The next obvious point that needs to be made explicit is the link between reliability and loss management. Line managers and operators can not keep costs down and revenues up without process reliability. And they can not lock on to reliability without a well-planned and consistently executed risk management program.

All of this is obvious, but if we do not make it explicit—if we do not declare that it is part of "the way we do things around here"—it will not happen.

The opposite is equally true. If we repeatedly and consistently declare that loss management is "the way we do things around here" then we begin to imbed it into the culture of the company. If we make it an explicit part of every business process, then we make it an approved tool for cost control. In this way, it rapidly becomes a routine part of work planning, business planning, performance monitoring and leadership activities.

Then everyone understands that you have not got effective control of your business until you have addressed loss management issues. Only then are people empowered to take the initiative to identify risks and seek resolutions.

At this point, loss management is part of everyone's responsibility—and everyone knows it.

■ PETRO-CANADA ACTION

By integrating loss management into the fabric of our business planning and operations we have found there are very quantifiable and substantial returns in all areas.

Environment—Consulting with the stakeholders on the earliest concepts of our plans for our MacKay River in situ oil sands development, we decided to use a Life Cycle Value Assessment approach. We calculated the costs of up-front environmental design improvements against estimated savings through the full life cycle of the facility and found some very attractive paybacks. We are all generally aware that up-front environmental design is far less expensive than retrofits. But we also found that it reduced operating costs.

In total, various design changes reduced projected greenhouse gas emissions by 150,000 tonnes, while water recycling reduced our water use

profile. At the same time we improved capital and operating costs by 7 million dollars, NPV. And the consultative process demonstrated good will and improved our environmental image. The hearing for our project went very smoothly. We received quick approval and we were able to break ground on schedule.

The Life Cycle Value Assessment process was so successful that we have currently applying it to four or five other projects, including our projects for the reduction of sulphur in gasoline.

Health—By taking an active consultative role in the management of our short-term disability cases, we were able to develop modified work programs for affected employees. As a result, we are bringing more employees back to work earlier and with better long-term results. In 1999 alone we estimated that this program saved us 2.7 million dollars.

Safety—In 1994, we changed the way we managed safety. We had previously assigned performance targets to work groups, with little or no impact on lost-time-incident rates. This time, we set targets, made interim checks, followed up on areas of concern and linked senior management's incentive compensation to improved safety performance. We achieved the desired issue focus and we have cut our recordable incident frequency in half and our lost time injury rate by one third.

Reliability—Over the past decade in the downstream, we have achieved significant reductions in total maintenance costs while increasing overall mechanical and operational availability. We also had a very positive trend in our reliability index—until we decided in 1998 that functional leadership and stewardship on reliability could be turned over to the plants.

Over the next two years, we experienced a decline in our overall reliability index from 88 percent to 78 percent, inadvertently proving our own point about the need for consistent executive leadership of these issues. By renewing our management focus in this area, we have moved our reliability index to 95 percent plus, so far this year.

■ NEW ROLE

I would submit that all of this changes the profile of loss management and the roles of the loss management professional, the line manager, the senior manager and the senior executive.

The loss management specialist is no longer an auditor, inspector and steward of loss management activities who stands outside the business unit. Instead, he or she is a consultant who provides line managers with the tools and training to achieve their loss management goals.

Functions of the loss management specialist include:

> maintaining a database on all losses;
> directing accident investigations and root-cause analyses for the business units; and,
> integrating all aspects of loss management, regardless of company structure. this role spans the traditional functions of health, safety, environment, reliability and process hazard management, plus related training and operating procedures.

The specialist also helps identify and integrate industry best practices into company work processes. In our own loss management system, a key source of new ideas has been our business alliance partners in the development of the Terra Nova offshore oil project. Our Terra Nova alliance has been able to cut injury rates to less than half of the norm for offshore construction work. What we have learned about safe project management from our Terra Nova alliance partners, we are now incorporating into planning for our oil sands business unit.

This new role of the loss management specialist presupposes that there are people ready, willing and authorized to use loss management as a tool for achieving business targets.

It requires that executives believe loss management is indispensable to reliability. It demands that executives are committed to building loss management into the planning and decision-making processes of the company. It requires that these executives and their senior managers are prepared to build loss management objectives into their own performance targets and compensation systems. It demands that they be willing to make their commitment to these issues visible and to communicate their commitment on a regular and sustained basis.

Senior managers must help form and maintain a network of loss management expertise within their areas of responsibility and across the business. They must also provide the resources to carry out programs and reach loss management objectives. One of the resources that is of most value to senior managers is their own time and if they are consistently seen to invest that resource in loss management priorities, it sends a very powerful message through the organization.

Finally, line managers must recognize loss management as a tool for meeting their own performance targets. This means line managers and team leaders must be trained and supported to act as coaches to employees, to raise employees' risk awareness and to assign individuals to do the ground work of implementing loss management principles and initiatives.

■ CONCLUSION

What I would suggest is that what you have at the end of this process is a new reality that looks very different than the old system of narrowly-defined safety policing.

This new reality recognizes that risk and loss management go beyond safety statistics to embrace everything from environmental design to process reliability. It comprehends that risk and loss management are tools for improving performance. And it builds these tools into the business by integrating loss management into planning and performance measurements.

For this to happen, executive leadership must understand the business imperative to demonstrate commitment through goal setting, communications, resource allocation and ongoing stewardship of the issue.

In this new reality, loss management leadership becomes part of the corporate vision. There is a higher level of ownership for the loss management issue at all levels of the company and it has become "part of the way we do things around here."

Increased use of risk and loss management tools leads to rising levels of internal expertise and measurable improvements in performance. And by performance I mean both reduced frequency of loss incidents of all kinds, as well as improved overall reliability and more stable revenues.

It is also our experience at Petro-Canada that, as we improved our risk and loss management practices, we increased public trust. As we are increasingly seen to be a responsible and responsive corporate citizen, this in turn increases our freedom of action. We have found that as project permitting has become more contentious, project design has become more consultative. Through this dynamic we have found there is a competitive advantage in demonstrating credibility and a proven loss management track record.

We have also found that our company is in a far better position to respond to every business opportunity and challenge because fewer surprises are hitting the bottom line and the company is competing at a high level of total fitness.

REFERENCES

Alberta Labour, Occupational Health and Safety Branch. *Occupational Health and Safety Manual for Small Business, 1990.*

American Institute of Chemical Engineers. *AIChE Technical Manual: Dow's Fire Explosion Index: Hazard Classification Guide, 1994.*

Association of Professional Engineers, Geologists and Geophysicists of Alberta (APEGGA). *Environmental Practice, A Guideline.*

Bird, Jr., F.E. and G.L. Germain. *Practical Loss Control Leadership.* Loss Control Management, Det Norske Veritas, 1992.

Canadian Standards Association. CAN/CSA-Q634-91. *Risk Analysis Requirements and Guidelines. Quality Management. A National Standard of Canada.*

Canadian Standards Association. *Emergency Planning for Industry.*

CCPS—American Institute of Chemical Engineers: *Guidelines for Hazards Evaluation Procedures.* 2nd edition, April 1995.

Construction Owners Association of Alberta. *The Owner's Role in Construction Safety—It Really Pays, 1991.*

Construction Owners Association of Alberta. *An Owner's Guide for a Contractor's Health and Safety Management Program, 1996.*

Energy Resources Conservation Board (ERCB). *Lodgepole Blowout Report,* 1984. [Note: The ERCB is currently called the Alberta Energy and Utility Board (AEUB).]

Engineering Council, the (UK). *Guidelines on Risk Issues.* L.R. Printing Services Limited, 1992.

Flixborough Disaster, The: Report of the Court Inquiry. Department of Employment, London, England (H.M. Stationary Office), 1975.

Kletz, T. *HAZOP and HAZAN.* 4th edition. IChemE, 1999.

Kletz, T. *What Went Wrong?* Gulf Publishing Company, 1994.

Marsh and McLennan. *Large Property Damage Losses in the Hydrocarbon-Chemical Industries: A Thirty-Year Review.* M&M Protection Consultants, 1995.

"Phillips 66 Explosion and Fire"; a report to the President, US Department of Labor, April 1990.

University of Alberta, Faculty of Engineering. *ENGG 404 Industrial Safety and Loss Management.* Industrial Safety and Loss Management Program.

University of Alberta, Faculty of Engineering. *ENGG 406 Industrial Safety and Risk Management.* Industrial Safety and Loss Management Program.